GEORGIA: CURRENT ISSUES AND HISTORICAL BACKGROUND

GEORGIA: CURRENT ISSUES AND HISTORICAL BACKGROUND

T.O. SIRAP (EDITOR)

Nova Science Publishers, Inc.
New York

Senior Editors: Susan Boriotti and Donna Dennis
Coordinating Editor: Tatiana Shohov
Office Manager: Annette Hellinger
Graphics: Wanda Serrano
Editorial Production: Jennifer Vogt, Matthew Kozlowski, Jonathan Rose,
                          Alexandra Columbus and Maya Columbus
Circulation: Ave Maria Gonzalez, Vera Popovich, Luis Aviles, Melissa Diaz,
                 Nicolas Miro and Jeannie Pappas
Communications and Acquisitions: Serge P. Shohov
Marketing: Cathy DeGregory

Library of Congress Cataloging-in-Publication Data
Available Upon Request

ISBN 1-59033-415-9.

Copyright © 2002 by Nova Science Publishers, Inc.
                    400 Oser Ave, Suite 1600
                    Hauppauge, New York 11788-3619
                    Tele. 631-231-7269   Fax 631-231-8175
                    e-mail: Novascience@earthlink.net
                    Web Site: http://www.novapublishers.com

Printed in the United States of America

CONTENTS

PREFACE

Since regaining its independence in 1991, Georgia has faced formidable problems of economic collapse, civil war, separatist conflict, rampant crime, political infighting and human rights abuses. This new book offers basic facts about today's Georgia but provides a detailed historical background of this beautiful, highly cultured country famous for its scholars, theater and hospitality among other virtues.

Chapter 1

GEORGIA: BASIC FACTS

Jim Nichol

SUMMARY

Since regaining its independence in 1991, Georgia has faced formidable problems of economic collapse, civil war, separatist conflict, rampant crime, and political infighting.

BACKGROUND

Land Area and Population

Area: 26,872 sq. mi.; slightly larger than the state of West Virginia. Population: 5.4 million (1995 est.). Administrative subdivisions include the Abkhazian Autonomous Republic and the Adzharian Autonomous Republic. The Georgian Supreme Soviet stripped South Ossetia of its status as an autonomous region (referred to by Georgians generally as the Samachablo region) in December 1990, following its persistent demands to secede and become a part of Russia.

History

Georgia has been subject to control or invasion by Romans, Persians, Arabs, Seljuk Turks, Mongols, and Ottoman Turks. At the beginning of the 19th century, Russia began its piecemeal annexation of Georgia, which was largely completed with the subjugation of the mountainous north-central area in 1858. Following the collapse of the Tsarist empire, Georgia declared its independence in 1918, and eventually secured diplomatic recognition from 22 countries. In February 1921, Russia's Red Army invaded and soon conquered Georgia. Georgia was included in the Transcaucasian Socialist Federal Soviet Republic (TSFSR), but after some areas of Georgia were ceded to other republics or states, the remainder became the Georgian Soviet Socialist Republic in 1936. Like many other non-Russian nationalities in the Soviet Union, Georgians can point to long-standing grievances against Soviet Communist rule that fueled their efforts to regain their independence. Many Georgians fell victim in the late 1920s and 1930s to collectivization, crash industrialization, and Stalin's purges (despite his Georgian-Ossetian ethnic origins). Nationalist riots were brutally suppressed in 1924 and 1956. Massive demonstrations took place in 1978 against Georgian government efforts to revise the Georgian constitution to give Russian and other languages equal status to Georgian. Increased popular discontent culminated in massive demonstrations in 1988 calling for democratization and respect for cultural monuments and the environment. In April 1989, many Georgian demonstrators were murdered, some with shovels, by Soviet military and police forces during a peaceful pro-independence march and protest against perceived Russian support for Abkhaz autonomy demands.

Georgia's first multiparty legislative elections were held in October 1990, resulting in a victory for the party coalition Round Table-Free Georgia, headed by academic and dissident Zviad Gamsakhurdia. He was subsequently selected by the deputies to serve as chairman of the legislature. On March 31, 1991, independence was endorsed by almost 90% of the electorate in a referendum, and a formal declaration of independence was unanimously approved by the legislature on April 9, 1991. Gamsakhurdia was popularly elected President of Georgia in May 1991. He nevertheless faced determined opposition from, among others, parties belonging to the National Congress, a national liberation body formed in October 1990. The Mkhedrioni paramilitary group, led by Jaba Ioseliani, was allied with the National Congress. During 1991, Gamsakhurdia's erratic attempts to remake Georgian society and politics triggered the resignations of his prime minister and foreign minister. The head of the National Guard, Tengiz Kitovani, also went into opposition.

The National Guard and Mkhedrioni spearheaded a general assault to overthrow Gamsakhurdia on December 21, 1991, forcing him to flee the country in early January 1992. A Military Council formed by Ioseliani, Kitovani, and others assumed power at that time, suspending the Soviet-era constitution (and substituting a 1921 constitution), dissolving the legislature, and declaring emergency rule. Former Georgian leader Eduard Shevardnadze was invited in early March 1992 to head a provisional government. He formed a civilian State Council (replacing the Military Council) to rule until elections could be held, and was elected head of its four-member presidium. The State Council was dissolved in October 1992 after legislative elections were held. In May 1993, Shevardnadze moved to consolidate his power by securing the resignations of Kitovani and Ioseliani from government posts. Gamsakhurdia returned from exile in September 1993 to the western Georgian region of Mingrelia and led a revolt to unseat Shevardnadze. Pro-Shevardnadze forces, assisted by the Russian military, were able to put down the revolt by early November 1993. Gamsakhurdia's death was reported in early January 1994. In further moves by Shevardnadze to consolidate power, Kitovani was arrested in January 1995 for planning an illegal paramilitary attack on Abkhazia, and efforts were being undertaken to disband Ioseliani's Mkhedrioni. On August 29, Shevardnadze survived a car bombing. He vowed to continue his anticrime efforts and announced his candidacy in upcoming presidential elections.

Republic Ethnic Composition and Tensions

3.8 million (70.1%) Georgian; 437,000 (8.1%) Armenian; 341,000 (6.3%) Russian; 308,000 (5.7%) Azerbaijani; 164,000 (3.0%) Ossetian; 100,000 (1.9%) Greek; 96,000 (1.8%) Abkhazian; and others (1989 census). Georgia is subject to ethnic and territorial disputes that threaten its territorial integrity, including separatism by Abkhazia and South Ossetia and dissident activities by other ethnic minorities. Georgia's armed conflict with South Ossetia escalated in December 1990, ending with a cease-fire agreement secured by Russian, Georgian, and Ossetial peacekeepers in June 1992. Though this cease-fire has been repeatedly extended, South Ossetia's demand to secede has not been resolved.

Massive and protracted armed conflict with Abkhazia began on July 23, 1992, when the Abkhaz Supreme Soviet declared its independence from Georgia. This prompted Georgia to send in troops in order to quell Abkhaz secessionist demands. The ensuing conflict led to thousands of casualties, hundreds of thousands of persons displaced, and the ruin of the central and regional economies. After 14 months of civil war, the Abkhazians were able to drive Georgian forces out of Abkhazia, thereby winning de facto control. The fall of Abkhazia led up to 250,000 ethnic Georgians and others to flee the region, creating a humanitarian crisis addressed by aid from several countries and international organizations. Following three failed attempts, a Russian-brokered cease-fire was signed by Georgian and Abkhazian delegations on May 14, 1994. As a provision of this agreement, an estimated 1,400-2,500 Russian troops (formally acting as Commonwealth of Independent States or CIS "peacekeepers") have been deployed in a security zone along the Inguri River which divides Abkhazia from Georgia proper. The U.N. Security Council in July 1994 approved an increase in the personnel of the U.N. military observer mission in Georgia (UNOMIG) to 136 to monitor the cease-fire and help refugees to return. The U.N. Secretary General reported in August 1995 that cooperation between UNOMIG and the CIS "peacekeepers" was satisfactory. He also planned to soon appoint a deputy to his special envoy to assist in detailed peace talks and to head UNOMIG, and called for establishing a U.N. human rights monitoring mission in Abkhazia to foster repatriation and the protection of human rights.

Large percentages of Armenians and Azeris in three southern Georgian regions have heightened concerns by some Georgians about the "demographic weakness" in these regions. Georgian societies such as Merab Kostava have encouraged Georgian migration into these areas. Some Georgians resent Armenians because of Armenians' economic dominance of Georgia in the 19th century. Religious differences between the Georgian Orthodox majority (65% of the population) and Islamic minority (11%) also contribute to political tensions. Georgia remains reticent to allow the repatriation of most Meskhetians, a Turkified and Islamized ethnic Georgian people who were deported during the Stalin period and who number 300,000 or more.

Political Leaders

Chairman of the State Council (legislature): Eduard Shevardnadze (head of state); Prime Minister: Otar Patsatsia; Defense Minister: Lt. Gen. Vardiko Nadibaidze; Foreign Minister: Aleksander Chikvaidze. Shevardnadze, born in 1928, rose through Komsomol (Communist Youth) ranks. After serving as district party leader in two cities, in 1964 he moved into police and security work, becoming Minister of Public Order (Internal Affairs) in 1965. In 1972, he became First Secretary of the Georgian Communist Party, charged with fighting corruption, increasing economic performance, and purging the entrenched leadership. In 1985, Gorbachev appointed him Minister of Foreign Affairs and he was promoted to full member of the ruling Soviet Communist Party Politburo. He effectively carried out the new conciliatory foreign policy termed "new thinking," earning significant criticism by conservative and Communist opponents of Gorbachev's policies. Faced with an apparent lack of support from Gorbachev and heightened criticism that he felt presaged a reactionary coup, he resigned in December 1990. He briefly reassumed the post of Soviet foreign minister in 1991 just before the demise of the Soviet Union.

Political Parties and Groups

Major political parties and organizations include the National Democratic Party (NDP), Georgia's Citizens Union (Shevardnadze's party), United Republican Party (URP), Union of Traditionalists (UT), Ilia Chavchavadze Society, United Communist Party (UCP, an umbrella party), Merab Kostava Society (MKS), Round Table, and the National Independence Party (NIP). Supporters of late president Gamsakhurdia comprise the Zviad Party and others. Most political parties and groups characterize themselves as in opposition to the Shevardnadze government. In June 1994, the Republican Party, the Popular Front, and Charter-91 merged, forming the URP. In anticipation of November 1995 legislative and presidential elections, new parties and blocs are forming, including an "Independent Georgia" bloc formed by the NIP, MKS, and forces allied with Ioseliani, and a bloc formed by the UT and Liberal Constitutional Party.

Political Institutions and Policies

A legislature of 235 members was elected on October 11, 1992, in multiparty voting generally adjudged "free and fair." Shevardnadze, who ran uncontested, was elected speaker of the legislature, gaining 95% of the popular vote. The October 1992 legislative elections heralded the creation of a political system where the legislative head served as the highest official and the presidency was abolished. After convening on November 4, the legislature, two days later, granted Chairman Shevardnadze wide-ranging powers as head of state pending completion of a new constitution. Work on a new constitution began in early 1993, and was finally approved by the legislature on August 24, 1995. The new constitution reestablishes a strong presidency, though it affirms a balance of executive and legislative powers more equitable than those reflected in most other new constitutions being approved by former Soviet republics. It calls for a unicameral, 235-member legislature. Federal and local government provisions were not included, however, with agreement that they would be added when the territorial integrity of Georgia has been assured. An electoral law was also approved for presidential and legislative elections to be held on November 5, 1995. Legislators will be elected in single-mandate constituencies and on the basis of party lists. A party must receive at least 5% of the party list vote to win legislative seats.

Human Rights and Freedoms

The human rights situation has been judged by many observers as poor due to widespread ethnic conflict, political turmoil, and crime. Freedom of assembly is restricted in practice, as is freedom of the press. Georgian police, paramilitary groups, the National Guard, and Security Service have engaged in widespread torture, beatings, and other actions against detainees to extract information or force confessions, according to the U.S. State Department and others. Many abuses have been committed against Gamsakhurdia supporters, or against Abkhazians and other ethnic minorities. Legal rights of detainees such as access to counsel have been violated. Georgian officials, on the other hand, accuse pro-Gamsakhurdia forces of myriad abuses. They also accuse Abkhazians of engaging in "ethnic cleansing" and atrocities against ethnic Georgians in Abkhazia. Police abuses have reportedly eased during 1994-1995 as Shevardnadze seeks to purge them of criminal and corrupt elements, and to disband paramilitary units. The Security Service, however, largely remained outside Shevardnadze's influence, under the control of the legislature and forces loyal to the UCP and Mkhedrioni, though the new constitution places it under presidential authority. Human rights organizations widely criticized the detention and sentencing in March 1995 of nineteen Gamsakhurdia supporters accused of an assassination attempt against Ioseliani.

Relations with Commonwealth of Independent States (CIS)

Both Gamsakhurdia and Shevardnadze had initially refused to join the CIS, citing its likely domination by Russia and its ineffectiveness. As part of the price for Russia's aid in defeating Gamsakhurdia's forces in 1993 and for otherwise taking a more conciliatory stance toward Georgia, on October 23, 1993, Georgia entered the CIS as its twelfth member. Shevardnadze further strengthened ties with Russia by signing a Friendship Treaty on February 3, 1994, which included a pledge to grant Russia the use of three military bases in Georgia. Georgia's entrance into the CIS was regarded as anathema by many Georgian nationalists and others, though Shevardnadze stressed the "realism" of accommodating Russia's interests. In March 1995, a new treaty went beyond the 1994 treaty by granting Russia rights to four military bases through the year 2020. Georgia has been unable to prevent repeated sabotaging of pipelines sending fuel to Armenia, leading to frictions in relations with Armenia. Azerbaijan and Armenia have also raised concerns about human rights conditions of ethnic Azeris and Armenians in Georgia. Georgia has refinanced a large debt owed for Turkmen natural gas.

Foreign Policy and Defense

Shevardnadze has pursued good relations with both East and West in pursuit of security and humanitarian assistance. Besides visits to former Soviet republics, in March 1994, he met with President Clinton in Washington, addressed the U.N., and visited Canada. Georgia has an ongoing interest in ties with the approximately one million Georgians residing in Turkey and the approximately 50,000 Georgians residing in Iran. Offices of the U.N. and the International Red Cross have been set up in Georgia and several delegations from the U.N., European Parliament, and the Organization on Security and Cooperation in Europe (OSCE) have traveled to Georgia, Abkhazia, and South Ossetia to examine the situation. Georgia is a member of the U.N., CSCE, NATO's Partnership for Peace, International Monetary Fund, World Bank, International Bank for Reconstruction and Development, and European Bank for Reconstruction and Development. Georgia's request for admission into the Council of Europe was deferred pending settlement of the Abkhazian conflict. Georgia is a member of the Black Sea Economic Cooperation Organization, headquartered in Turkey.

To help remedy the chaotic state of Georgia's embryonic armed forces, Shevardnadze decreed in late May 1993 (and the legislature later approved) setting up a Soviet-style "single, well-armed army." A military doctrine and program were worked out for eventually creating a volunteer army of 15,000. The 1994 Friendship Treaty and agreements signed in March 1995 call for Russian aid in equipping and training the army. Although the Georgian legislature had called for the total withdrawal of Russian troops, the deployment of Russian CIS "peacekeepers," under the terms of the May 1994 Georgian-Abkhazian cease-fire, as well as the use of military bases in Georgia by Russian military forces, made this goal moot. Shevardnadze reportedly told visiting U.S. Chairman of the Joint Chiefs of Staff John Shalikashvili in May 1995, who offered some military technical aid, that Georgia's close military ties with Russia were dictated by Georgia's paramount goal of resolving Abkhaz separatism. NATO's Deputy Supreme Allied Commander visited Georgia in late July 1995 to plan Georgia's participation in PFP.

Economy

Georgia's GDP in 1994 was estimated at $6.3 billion and per capita income at $1,164 (Planecon data; purchasing power parity rate, 1990 dollars). This represents a drop from $9 billion GDP and $1,662 per capita income the previous year. The Georgian economy continued to be in crisis; industrial production fell 40% and agricultural production fell 10% in 1994, with dozens of major companies shutting down. The value of the currency, the coupon, plunged as producer prices increased 12,000% in 1994, though inflation has eased during 1995. There was also some reported success in raising state revenues from taxes, though government expenditures still outpaced revenues by some 75% in the first half of 1995. About 40-60% of agricultural land and some housing were privatized during 1992, but few industries or retail outlets. Turkey, Georgia, and Azerbaijan are studying upgrading a Georgian oil pipeline to transport some Azeri oil to Black Sea ports.

U.S. POLICY

The United States recognized Georgian independence on December 25, 1991, but delayed establishing diplomatic relations, citing a poor human rights record and civil disorder. The United States moved to extend humanitarian aid upon Shevardnadze's return to Georgia in 1992; diplomatic relations with Georgia were established in late April of that year. The United States has urged the peaceful settlement of the Abkhaz conflict, discussing the conflict at the April 1993 U.S.-Russian Summit Meeting and within the U.N. Security Council. During Shevardnadze's state visit to the United States in March 1994, President Clinton "reaffirmed in very strong terms America's support for the independence, the sovereignty, and the territorial integrity of Georgia." Through the end of March 1995, the United States has obligated cumulative aid of $435.5 million to Georgia, mostly for emergency relief. Georgia ranked second among the newly independent states of the former Soviet Union, behind Armenia, in U.S. aid per capita. The Administration has requested $21 million in assistance to Georgia for FY1996, with $12 million planned for humanitarian relief.

Chapter 2

GEORGIA [REPUBLIC]: CURRENT DEVELOPMENTS AND U.S. INTERESTS[1]

Jim Nichol

SUMMARY

Since regaining its independence in 1991, Georgia has struggled to surmount economic problems, civil war, crime, political infighting, and human rights problems. Following the September 2001 terrorist attacks on the United States, Georgia's leaders have offered support to the U.S.-led anti-terrorism campaign in Afghanistan. Tensions between Georgia and Russia have increased, however, as Russia has intensified anti-insurgency efforts in Chechnya, neighboring Georgia. Some sporadic fighting has resumed in the breakaway republic of Abkhazia. In response to anti-government protests, President Shevardnadze dismissed the entire cabinet of ministers in November 2001.

---

[1] Sources include Foreign Broadcast Information Service, *Daily Report: Central Eurasia*; Radio Free Europe/Radio Liberty, *Newsline*; the Economist Intelligence Unit; Jamestown Foundation, *Daily Report and Prism*; *World Bank*; *International Monetary Fund*; State Department and U.N. information; and Reuters and Associated Press wire service reports.

U.S. POLICY

U.S.-Georgian relations have been close, reflecting U.S. recognition of President Eduard Shevardnadze's former role as a pro-Western Soviet Foreign minister, his current pro-Western foreign policy, Georgia's democratization efforts, and its importance as an east-west transit route, including for oil and natural gas pipelines. The United States has supported efforts to reach peaceful settlements to the ethnic conflicts in Georgia by the United Nations and Organization for Security and Cooperation in Europe (OSCE), both of which maintain observer missions in Georgia. The United States has designated a Special Negotiator for Regional Conflicts in the NIS to assist in these efforts. Shevardnadze visited the United States on several occasions during the Clinton presidency. He met with President Bush and other Administration officials and Members of Congress in Washington in October 2001.

The United States has been Georgia's largest bilateral aid donor, obligating cumulative aid FY1992-FY2000 of $787.6 million. Shevardnadze has often stated that U.S. humanitarian aid made a critical difference in averting famine in Georgia in the early 1990s. Other U.S. aid has been used for training judicial and law enforcement personnel, enhancing border and export controls, privatization, reforming the tax code and budgetary process, building the legal basis for foreign investment in the energy sector, training for military personnel, and providing NATO-compatible defense equipment so Georgia can more fully participate in NATO's Partnership for Peace (PFP) activities. The Defense Department (DOD) also provides excess commodities and privately donated aid transported by DOD, not included in cumulative obligations. Estimated U.S. aid to Georgia for FY2001 is $98.9 million. The United States also provides funds for U.N. observer activities in Georgia.

Georgia: Basic Facts

Area and Population: 26,872 sq. mi., slightly larger than West Virginia. The population is 5.5 million (2000 est., *World Bank*). Administrative subdivisions include the Abkhazian and Ajarian Autonomous Republics. The Georgian Supreme Soviet stripped South Ossetia of its autonomous status in late 1990, following its demands to secede and become a part of Russia.

Ethnicity: 70.1% are Georgian; 8.1% Armenian; 6.3% Russian; 5.7% Azerbaijani; 3.0% Ossetian; 1.9% Greek; 1.8% Abkhazian, and others (1989 census).

GDP: $3.0 billion in 2000 (*World Bank* est.).

Language and Religion: Most Georgians speak Caucausian languages, though some speak Iranian or Turkic languages. Most Georgians and Ossetians are Orthodox Christians, and many Ajarians and some Abkhazians are Sunni Muslims.

Independence: The Georgian legislature approved a declaration of independence on April 9, 1991. U.S. recognition occurred on December 25, 1991, and diplomatic ties were established in April 1992.

Congressional interest in Georgia has been demonstrated by earmarks on assistance in yearly appropriations, creation of a South Caucasus funding category to be used to pursue confidence-building and other conflict resolution measures "in the vicinity of Abkhazia" and elsewhere, creation of a Georgian border security fund, and in provisions in the Security Assistance Act of 2000 (H.R. 4919; P.L. 106-280; signed into law on October 6, 2000). The latter authorized $45.5 million over two years to assist Armenia and the GUUAM countries (a group formed in 1997 and named after the initials of the members – Georgia, Ukraine, Uzbekistan, Azerbaijan, and Moldova) to strengthen national control of borders and to promote independence and territorial sovereignty.

In FY2002 appropriations bills, Congress has earmarked $90 million in Freedom Support Act assistance for Georgia (H.R. 2506). Appropriators expressed concerns about corruption in Georgia and the limited progress by Russia in closing its military bases in Georgia.

POLITICAL AND ECONOMIC DEVELOPMENTS

Political Parties and Groups

Major political parties that won representation in the legislature in November 1999 include Shevardnadze's ruling Citizens' Union of Georgia party (gaining 891,000 of 2.1 million party list votes cast), Ajarian leader Abashidze's Revival Union (537,000 votes), and Industry Will Save Georgia (151,000 votes). The Georgian Labor Party just failed to gain enough votes to win party list seats (141,000 votes). Other parties that gained more than one percent of the party list vote include the opposition National Democratic Party (NDP), the People's Party, and the United Communist Party.

Political Institutions and Policies

Until 1995, Georgia was governed according to a constitution dating back to the early Soviet era. In August 1995, parliament approved of a new constitution giving the presidency added powers. The new constitution established a unicameral, 235-member legislature elected by single-mandate constituencies (85 seats) and party lists (150 seats). Government ministers are responsible to the president, who is assisted by a state minister. Shevardnadze in December 1999 decreed enhanced powers for the state minister "equal to those of a prime minister." The speaker's only constitutional powers are to sign bills and serve as acting president in case the president is indisposed or dies. The legislature agreed that federal provisions would be added to the constitution after Georgia's territorial integrity has been assured. In 2001, Shevardnadze promoted a constitutional amendment to re-create a post of prime minister, but parliament failed to act on the measure.

The OSCE reported that legislative races in October-November 1999 in Georgia appeared mostly fair, but irregularities in second-round voting marred full compliance with OSCE standards. Thirty-two parties and blocs were registered. Seven candidates registered to run in Georgia's April 9, 2000, presidential election. The Georgian Central Electoral Commission reported that Shevardnadze received 80% of 1.87 million votes. OSCE monitors reported that the government aided the incumbent, state media were biased, vote counting and tabulation procedures lacked full transparency, and ballot box stuffing had taken place, so that the election did not meet OSCE standards, though "fundamental freedoms were generally respected during the election campaign and candidates were able to express their views." Although healthy, Shevardnadze's age (73), previous threats to his life, and growing political isolation have fueled speculation about possible successors.

On November 1, 2001, President Shevardnadze dismissed his entire cabinet in response to anti-government rallies following a police raid on an independent television station. He did not heed some protesters' calls for his own resignation. Underlying reasons for the sacking may have been internal divisions inside the ruling CUB party between reformers and hardliners. Shevardnadze put forth a list of candidates for a new cabinet on November 15. Early elections may be called.

Political Leadership

President: Eduard Shevardnadze. Shevardnadze dismissed the cabinet of ministers on Nov. 1, 2001. Parliament elected Nino Burjanadze to be Speaker of the Parliament on Nov. 10. On Nov. 14, Shevardnadze proposed a new cabinet that includes many members of the previous government. The next parliamentary elections are scheduled to be held in late 2003, and the next presidential vote is in April 2005.

Biography: Shevardnadze, born in 1928, began police and security work in Georgia in 1964. In 1972, he became First Secretary of the Georgian Communist Party. In 1985, then-Soviet leader Mikhail Gorbachev appointed him Foreign Minister, to carry out the conciliatory foreign policy termed "new thinking." Increasingly reproached by hardliners and feeling that Gorbachev was not supporting him, Shevardnadze resigned in late 1990, though he briefly re-assumed the post in 1991. He was elected Georgia's legislative speaker in 1992 and president in 1995 and 2000. He has said he will not seek another term.

Human Rights and Freedoms

According to the U.S. State Department's *Country Reports on Human Rights Practices for 2000*, the Georgian government's human rights record worsened in the past year. The report cited serious irregularities in the April 2000 presidential election, as well as continued police abuses and problems with the criminal justice system. It called corruption in enforcement agencies "significant and pervasive." About 200 independent newspapers and some free electronic media operate, though some harassment cases were reported. Freedom of assembly is generally upheld. Independent non-governmental organizations have actively defended citizens' rights. In October 2001, a police raid against an independent television station led to angry street demonstrations against the government, prompting Shevardnadze to sack the government, but not step down himself.

Defense

According to *The Military Balance 2000-2001*, Georgia's ground forces and a small navy and air force, and Ministry of Defense staff number about 26,900. There are about 5,000 Russian troops and 1,500 Russian "peacekeepers" in Georgia. Marking the shift toward more security ties with the West, Georgia withdrew from the CIS collective Security Treaty in early 1999. In January 1999, Georgia assumed full control over guarding its sea borders, and on October 15, 1999, the last Russian border troops (except some liaison officers) left Georgia. Georgian military officers have received training in Ukraine, Turkey, and the United States, and participate in NATO's PFP exercises. Western educated Defense Minister Tevzadze advocates reorganizing the armed forces into a "U.S. model" professional army, and Shevardnadze has called for them to be ready to join NATO by 2005. An ongoing problem is limited resources for the defense budget, which contributes to low pay and provisioning of troops and high rates of desertion.

The Economy

Georgia's economy has been set back by the loss of many of its markets in the former Soviet Union and by the disruption caused by ethnic and civil conflict. The severe decline in Gross Domestic Product (GDP) – some 80% over the period 1990-1994 – halted in 1995. The World Bank reported GDP growth of 3.0% in 1999, but only 1.8% in 2000. Budget deficits were around 4-4.5% in 1995-1999. These problems contributed in late 1998 to the devaluation of Georgia's currency, the lari. Almost all housing and most small enterprises have been privatized, and over 80% of agricultural holdings and medium and large enterprises, though the slow registration of land titles has held back progress. Some port facilities and resource industries have been designated as strategic, with the state retaining majority ownership. Widespread poverty and ethnic conflict have contributed to the emigration of about one-fifth (one million) of the population since 1991. The health care system, beset by budgetary shortfalls, battles rising infant mortality and communicable disease rates. Drought in 2000 harmed agriculture and led to international appeals for food aid.

Energy Issues

Georgia does not possess oil and natural gas deposits sufficient to meet its domestic energy needs, so it depends on imports (some recent exploration indicates added oil reserves), especially from Russia. An "early oil" pipeline opened in April 1999 that transits Georgia, delivering about 100,000 barrels of oil per day from Azerbaijan to Georgia's Black Sea port of Supsa (near Poti). Georgia gains sizable transit fees from this pipeline. Georgia is a proposed route for a "main oil" pipeline (carrying larger volumes of oil) to the Turkish port of Ceyhan, and for a gas pipeline. These pipelines will transport Azerbaijani, Kazakh, and Turkmen oil and gas, giving these states and Turkey stakes in Georgia's future stability. On November 18, 1999, Azerbaijan, Georgia, Turkey, and Kazakhstan signed the "Istanbul Protocol" on construction of a trans-Caucasus oil pipeline from Azerbaijan to Turkey (expected to be completed in 2004 with a capacity of one million barrels per day). In early 2000, Georgia and Azerbaijan settled terms for transit fees for the proposed oil pipeline. In September 2001, the two countries signed a second agreement on the transit, transportation and sale of natural gas from Azerbaijan through Georgia.

Relations with Russia and Other Countries

Georgia's relations with Russia have been uneasy. Though it has attempted to maintain working political, economic, and security ties with Russia, it seeks other ties with the West and international organizations to maximize its independence. Russia's objectives toward Georgia include retaining some influence in the South Caucasus region. Relations with Russia worsened in late 1999, after it launched new military attacks against its breakaway Chechnya region, which borders Georgia. Some fighting and refugees spilled over Georgia's borders, and Russia has regularly accused Georgia of aiding and abetting Chechen "terrorists" near to the border. The OSCE maintains an observer mission along Georgia's border with Chechnya, to help prevent the spillover of the conflict. Following the September 2001 terrorist attacks on the United States, Russia has alleged links between Osama bin Laden and the Chechen rebels. Some Russian officials have threatened military action in Georgia in order to eliminate the Chechen rebel threat. Some

Chechen forces have reportedly joined forces with Georgian guerrillas to attack the breakaway republic of Abkhazia. The Georgian parliament voted to demand the withdrawal from Abkhazia of Russian peacekeeping troops, whom parliament accuses of supporting the Abkhazian separatists.

At Shevardnadze's request in 1993, Russia's military helped defeat an insurgency aimed at overthrowing him, and in return, Georgia entered the Commonwealth of Independent States (CIS) and signed a CIS Collective Security Treaty. This was regarded as anathema by many Georgians, though Shevardnadze stressed the "realism" of accommodating Russia at that time. In March 1995, the two sides signed a treaty granting Russia rights to four military bases through the year 2020. The OSCE approved an adapted Conventional Armed Forces in Europe Treaty in Istanbul in November 1999, with Russia and Georgia agreeing that Russia would reduce its number of tanks in Georgia to not more than 53, armored combat vehicles to not more that 241, and artillery systems to not more than 140. After reported heavy U.S. lobbying, Russia also declared that it would close its Georgian military bases at Gudauta and Vaziani by July 1, 2001, and discuss with Georgia the disposition of two other bases. Russia turned over the Vaziani base by July 1, but not the Gudauta base, located in the breakaway region of Abkhazia. In November, Russia announced the withdrawal of its forces and equipment from the Gudauta base. However, the Georgian foreign ministry has questioned some of the terms of closure, such as the continued presence of some 600 Russian troops on the base.

Georgia's reform progress was marked by its admission to the Council of Europe in April 1999 and the World Trade Organization in October 1999. Among its neighbors, Georgia has good – though not problem-free – relations with Armenia and Azerbaijan. Georgia has an ongoing interest in ties with about one million Georgians residing in Turkey and about 50,000 Georgians residing in Iran. The GUUAM states share common interests in resisting Russian domination and in securing energy transport and supplies that are outside Russian control. In November 2001, GUUAM foreign ministers issued a joint statement on terrorism pledging to develop and reinforce cooperation with each other and with international organizations in order to combat terrorism.

Ethnic and Regional Tensions

Ethnic disputes compromise Georgia's territorial integrity. South Ossetians in 1989 called for joining their territory with North Ossetia in Russia or for independence. In June 1992, Russia's then-President Boris Yeltsin brokered a cease-fire, and a predominantly Russian military "peacekeeping" force of about 500 has since been stationed in South Ossetia. A coordinating commission to settle the conflict meets regularly, composed of OSCE, Russian, Georgian, and North and South Ossetian emissaries, but rapprochement remains elusive. Ajaria also practices substantial self-rule. The northwestern Mingrelia area has witnessed anti-government violence.

Abkhazia

In the early 1990s, the Abkhaz conflict resulted in about 10,000 deaths and over 200,000 refugees and displaced persons, mostly ethnic Georgians. In July 1992, the Abkhaz Supreme Soviet declared its independence from Georgia. This prompted Georgian national guardsmen to attack Abkhazia. In October 1992, the U.N. Security Council approved the first U.N. observer mission to a NIS state, termed UNOMIG.[2] In September 1993, Russian and North Caucasian "volunteer" troops that reportedly made up the bulk of Abkhaz separatist forces broke a cease-fire and quickly routed Georgian forces. In April 1994, the two sides signed framework accords on a political settlement and on the return of refugees and displaced persons. The next month, Georgia and Abkhazia signed a cease-fire providing for Russian troops (acting as CIS "peacekeepers") to be deployed in a security zone dividing Abkhazia from the rest of Georgia.

The U.S. Special Negotiator for Regional Conflicts works with the U.N. Secretary General, his Special Representative, and other Friends of Georgia (France, Germany, Russia, the United Kingdom, and Ukraine) to facilitate a peace settlement. Some U.S. personnel serve in both the U.N. and OSCE missions. In late 1997, the sides agreed to set up a Coordinating Council to discuss cease-fire maintenance and refugee, economic, and humanitarian issues. Meetings of the Coordination Council took place in 2000 and some confidence-building measures were agreed upon, but the U.N. Security Council noted a seeming lack of commitment by the parties to a comprehensive peace settlement. Several scheduled meetings of the Council in 2001 have been cancelled.

---

[2] In July 2001, the U.N. Security Council extended UNOMIG's mandate for another six months.

After a lull of several years, an upsurge in fighting in October 2001 in the Kodori Gorge region of Abkhazia has resulted in approximately 40 deaths. A U.N. helicopter was shot down by a missile on October 8, killing all nine persons on board. Abkhaz officials claim that Georgian guerrillas have joined forces with at least 200 Chechen rebels. Georgian officials deny any direct involvement in the fighting, but mobilized the armed forces to the Abkhaz border. The clashes have revived the threat of renewed violent conflict between Georgian forces and Abkhaz separatists backed by Russia.

Chapter 3

OIL AND NATURAL GAS IN GEORGIA

Joseph P. Riva, Jr.

SUMMARY

Georgia depends upon imports for most of its fossil energy needs. It produces only about 3,590 barrels of oil per day, while consuming 135,000 barrels. Georgia's proved oil reserves, an estimated 13 million barrels, are being efficiently produced. Since every drop of oil is needed, not even very small fields are shut-in. The known oil fields may have the potential to produce up to 100 million barrels, with the use of modern recovery technology, while undiscovered oil resources may range to twice that amount. Georgia annually produces about 1.4 billion cubic feet of natural gas, only about seven percent of its annual consumption. Natural gas reserves are an estimated 14 billion cubic feet, while up to 300 billion cubic feet may remain undiscovered.

INTRODUCTION

Georgia borders Armenia, Russia, Azerbaijan, Turkey, and the Black Sea. Its capital is Tbilisi. It is slightly larger than West Virginia, 26,872 square miles, and its population of 5.4 million is 69 percent Georgian, nine percent Armenian, five percent Azeri, three percent South Ossetian, two percent Greek, two percent Abkhazian, and 10 percent other nationalities. Russia's conquest of Georgia began in about 1800 and was not completed until 1878. However, Georgia gained its independence from the Tsarist Empire and established an independent republic in 1918. Although previously recognized by Soviet Russia, the Red Army invaded and conquered the country in 1921, under orders from Georgian Joseph Stalin. Georgia, Armenia, and Azerbaijan were united into the Transcaucasian Soviet Federal Socialist Republic from 1922 to 1936, at which time they were re-established as three separate republics within the Soviet Union. Despite Stalin's Georgian heritage, the region suffered greatly under Soviet rule.

Later, Gorbachev's more tolerant policies allowed nationalistic sentiments again to be expressed and the drive for Georgian independence was renewed. Multiparty elections were held in 1990 and independence was endorsed by nearly 90 percent of the electorate in a referendum in 1991. A month later, a president was elected with 87 percent of the popular vote. The president, however, was soon accused of attempting to establish a dictatorship and opposition forces revolted in Tbilisi. After heavy fighting that destroyed parts of the city, the president was expelled and a provisional government set up in 1992. As Georgia was moving toward independence from the Soviet Union, the Ossetians and Abkhazians were attempting to secede from Georgia. The Abkazinas want independence and have declared themselves a sovereign republic. The declaration has been annulled by Georgia, which has intensified its control over the Abkhazian region.

The South Ossetians, who supported Soviet Russia in its conquest of Georgia in 1921, reside in a 1,600 square mile region along the border with Russia. They hope to unite with Russia's North Ossetians and become a territorial unit within the Russian Federation. Fighting began late in 1990 between Georgian nationalist guerrillas and Ossetian rebels. Russia's policies toward the conflict are contradictory, reflecting its own internal power struggles. Although sending weapons to the Georgian government to help end the fighting, there also has been some Russian support for the South Ossetian separatists and for the recent closing of a gas pipeline to Georgia by the North Ossetians.

The Georgian economy revolves around fruits and teas, which generally are not grown elsewhere in the former Soviet Union. It has some tourism along the Black Sea coasts. There are significant coal deposits. Georgia has banned the export of all foodstuffs and has made little progress to privatization or a free market system.

CURRENT OIL AND GAS STATUS

Georgia is dependent on imports for most of its fossil energy requirements. Daily, it consumes some 135,000 barrels of oil while producing only about 3,590 barrels per day. It annually consumes 208 billion cubic feet of natural gas and produces only 1.4 billion cubic feet (see table 1). Georgia has only one oil refinery, with a capacity of 120,000 barrels per day. It does not have the capability to upgrade residual oil into lighter transportation products. Georgia's proved oil reserves are estimated at about 13 million barrels, with growth potential to perhaps 100 million barrels.[3] Undiscovered recoverable oil resources may range to about twice that amount. Natural gas reserves are estimated at about 14 billion cubic feet, while up to 300 billion cubic feet of gas may remain undiscovered. There have been plans to create a Europe-Asia oil transport corridor through Georgia from Kazakhstan to the Black Sea. The main condition for the project would be political stability in Georgia.

PETROLEUM GEOLOGY AND PETROLEUM DEVELOPMENT

Rioni Basin

The Rioni basin is located at the southeastern end of the Black Sea in Georgia. Near the coast, the contained Cretaceous and Cenozoic sedimentary sequence contains many oil shows.[4] Georgia lies within the Alpine-Himalayan fold belt system, formed by the convergence of the African-Arabian and Eurasian plates. However, basin development is complex and possibly dates to the Paleozoic, with basin closure and rejuvenation evident throughout its history (see table 2 for geologic time scale). During the early Tertiary, when the reservoir rocks, source rocks, and cap rocks were deposited, and when the traps were created, the basin was in a back-arc environment, marginal to active volcanism.[5]

---

[3] George, Dev. Redefining the Middle East: New Countries, New Politics, New Activity. Offshore, Mar. 1992. p. 25-27.

[4] Eremenko, N.A., Ya. P. Malovitskiy, I.S. Gramberg, and L.I. Lebedev. Geological Structure and Oil and Gas Prospects of USSR Continental Shelf. The American Association of Petroleum Geologists Bulletin, Feb. 1973. p. 235-243.

[5] Patton, D.K. Samgori Field, Republic of Georgia: Critical Review of Island-Arc Oil and Gas. Journal of Petroleum Geology, Apr. 1993. p. 153-168.

Table 1. Georgia Petroleum Status

Oil (million barrels)

Cumulative Production	200
Proved Reserves	13
Inferred Reserves	87
Undiscovered Resources (range)	100 (50-200)
Total Oil Endowment	400
1992 Production	1
Reserves/Production (R/P) Ratio	13/1
Potential Production at R/P – 9/1	1

Natural Gas (billion cubic feet)

Cumulative Production	6
Proved and Inferred Reserves	14
Undiscovered Resources (range)	100 (0-300)
Total Gas Endowment	120
1992 Production	1
Reserves/Production (R/P) Ratio	14/1
Potential Production at R/P = 9/1	2

Source: Derived from U.S. Geological Survey, Oil and Gas Journal, and Offshore.

Clay-rich rocks of the Paleocene-lower Eocene, upper Eocene, and Oligocene are potential hydrocarbon source rocks, provided their thermal maturity is suitable. However, the thermally-mature lower Eocene clay sediments are the most likely source of the oil in the basin. The deeper Mesozoic section may have been gas prone. Reservoir rocks include sandstones in the upper and lower Eocene and altered volcanic tuffs of middle Eocene age.[6] Most of the oil produced in the basin is from the middle Eocene hydrothermally altered tuffs, sealed by upper Eocene marls in anticlinal structures. The middle Eocene tuff has an andesite-basalt composition, and ranges between 800 and 1,850 feet in thickness. Porosity and permeability, however, are variable and influenced by fracturing and alteration.

---

[6] Ibid., and Grynberg, M.E., D. Papava, M. Shengelia, A. Takaishvili, A. Nanadze, and D.K. Patton. Petrophysical Characteristics of the Middle Eocene Laumontite Tuff Reservoir, Samgori Field, Republic of Georgia, Journal of Petroleum Geology, July 1993. p. 313-322.

Table 2. Geologic Time Scale
(millions of years)

Era	Period	Epoch	*Major Event*	Began	Duration
Cenozoic					
	Quaternary				
		Holocene	*Humans abundant*	0.01	0.01
		Pleistocene	*Humans appear*	1.6	1.59
	Tertiary				
		Pliocene	*Mammals diversity,*	5.3	3.7
		Miocene	*Grasses spread*	23.7	18.4
		Oligocene		36.6	12.9
		Eocene	*Mammals develop rapidly*	57.8	21.2
		Paleocene		66.4	8.6
Mesozoic					
	Cretaceous		*Dinosaurs become extinct, Flowering plants appear*	144	77.6
	Jurassic		*Birds appear*	208	64
	Triassic		*Mammals appear, Dinosaurs appear*	245	37
Paleozoic					
	Permian		*Reptiles appear*	286	41
	Carboniferous				
	Pennsylvanian		*Insects abundant*	320	34
	Mississippian			360	40
	Devonian		*Fish abundant*	408	48
	Silurian		*Amphibians appear Land plants and animals appear*	438	30
	Ordovician		*Fish appear*	505	67
	Cambrian		*Marine invertebrates abundant*	570	65
Precambrian					
	Proterozoic		*Simple marine organisms*	2500	1930
	Archean			(+/-)3800	(+/-)1300

Source: Derived from the Geological Society of America Time Scale, 1983.

Oil exploration began in Georgia in 1930. Between 1930 and 1973 some 450 exploration wells were drilled with disappointing results. In 1970, yearly oil production was less than 200,000 barrels and cumulative production to 1973 totalled only 8.4 million barrels. A surface anticline was mapped at *Samgori*, near Tbilisi, in 1965, with exploratory drilling beginning in 1967. The *Samgori* field was discovered in 1974 and since has produced more than 166 million barrels of oil, mostly from a middle Eocene andesite-basalt tuff that has been fractured and hydrothermally altered. The secondary porosity and permeability developed by fracturing and hydrothermal alteration is critical to the occurrence of adequate reservoir rock in the field. Some 130 wells have been drilled in the field, 117 of which have produced oil. Annual production peaked at about 21 million barrels in the late 1970s, and pumps were installed in 1986 following a steep production decline. Recently, about 50 wells were producing a combined yearly output of 872,000 barrels of oil. A deeper gas zone was discovered in 1990, and now about 585 million cubic feet of natural gas is being produced annually. The *Samgori* field occurs in a 15 mile long by two mile wide east-west trending anticline. The structure is asymmetrical, with the steeper dips on the southern limb. The source rocks are most likely Paleocene-lower Eocene clays.[7] Several other oil fields have been discovered in Georgia, and are included in table 3. *Rustavi* also produces about 778 million cubic feet of gas annually.

Table 3. Georgia's Oil Fields
(in 1,000 barrels)

Field (date discovered)	Annual Production	Cumulative Production
Samgori (1974)	872.0	166,000
Mirzaani (1930)	47.6	7,792
Patara-Shiraki (1932)	2.8	527
Noiro (1939)	14.3	1,799
Supsa/Shromisubani (1939)	19.5	789
Satskhenesi (1956)	7.6	320
Taribani (1963)	3.2	366
Chaladidi (1971)	1.4	142
Teleti (1977)	165.0	2,957
Southern Dome (1979)	160.1	7,873
Rustavi (1983)	13.5	26

Source: Journal of Petroleum Geology.

Most of Georgia's fields are very small. *Samgori* originally contained over 150 million barrels of oil, but the rest of the fields are less than one-tenth of *Samgori's* size. Annual production from these fields is very low. However, Georgia needs all of the oil that it can get so even fields producing only a few

---

[7] Patton, D.K., op. cit.

barrels per day are not shut-in. The existence of undiscovered larger petroleum fields in Georgia depends mostly on the occurrence of good reservoir rocks. Volcanic tuffs offer potential reservoirs, but only if their porosity and permeability is enhanced by fracturing and hydrothermal alteration. Since meteoric water has infiltrated the volcanic section over large areas, undiscovered reservoirs similar to *Samgori's* may exist. Exploration success would depend upon locating such fractured and altered portions of the volcanic rocks.[8] There is a projected 100 million barrels of undiscovered recoverable oil remaining in the Rioni basin within a range of 50 to 200 million barrels. Reserves are estimated at about 100 million barrels, with perhaps twice that amount already having been produced.

---

[8] Ibid.

Chapter 4

GEORGIA: A COUNTRY STUDY[*]

Glenn E. Curtis (Editor)

PREFACE

At the end of 1991, the formal liquidation of the Soviet Union was the surprisingly swift result of partially hidden decrepitude and centrifugal forces within that empire. Of the fifteen "new" states that emerged from the process, many had been independent political entities at some time in the past. Aside from their coverage in the 1989 *Soviet Union: A Country Study*, none had received individual treatment in this series, however. *Armenia, Azerbaijan, and Georgia: Country Studies* is the first in a new subseries describing the fifteen post-Soviet republics, both as they existed before and during the Soviet era and as they have developed since 1991. This volume covers Armenia, Azerbaijan, and Georgia, the three small nations grouped around the Caucasus mountain range east of the Black Sea.

The marked relaxation of information restrictions, which began in the late 1980s and accelerated after 1991, allows the reporting of nearly complete data on every aspect of life in the three countries. Scholarly articles and periodical reports have been especially helpful in accounting for the years of independence in the 1990s. The authors have described the historical, political, and social backgrounds of the countries as the background for their current portraits. In each case, the authors' goal was to provide a compact, accessible, and objective treatment of five

---

[*] Excerpted from the Library of Congress Federal Research Division Website.

main topics: historical background, the society and its environment, the economy, government and politics, and national security.

In all cases, personal names have been transliterated from the vernacular languages according to standard practice. Placenames are rendered in the form approved by the United States Board on Geographic Names, when available. Because in many cases the board had not yet applied vernacular tables in transliterating official place-names at the time of printing, the most recent Soviet-era forms have been used in this volume. Conventional international variants, such as Moscow, are used when appropriate. Organizations commonly known by their acronyms (such as IMF--International Monetary Fund) are introduced by their full names.

Autonomous republics and autonomous regions, such as the Nakhichevan Autonomous Republic, the Ajarian Autonomous Republic, and the Abkhazian Autonomous Republic, are introduced in their full form (before 1991 these also included the phrase "Soviet socialist"), and subsequently referred to by shorter forms (Nakhichevan, Ajaria, and Abkhazia, respectively).

Measurements are given in the metric system; a conversion table is provided in the Appendix. A chronology is provided at the end of the chapter, combining significant historical events of the three countries. To amplify points in the text of the chapters, tables provide statistics on aspects of the societies and the economies of the countries.

INTRODUCTION

The three republics of Transcaucasia--Armenia, Azerbaijan, and Georgia--were included in the Soviet Union in the early 1920s after their inhabitants had passed through long and varied periods as separate nations and as parts of neighboring empires, most recently the Russian Empire. By the time the Soviet Union dissolved at the end of 1991, the three republics had regained their independence, but their economic weakness and the turmoil surrounding them jeopardized that independence almost immediately. By 1994 Russia had regained substantial influence in the region by arbitrating disputes and by judiciously inserting peacekeeping troops. Geographically isolated, the three nations gained some Western economic support in the early 1990s, but in 1994 the leaders of all three asserted that national survival depended chiefly on diverting resources from military applications to restructuring economic and social institutions.

Figure 1. Armenia, Azerbaijan, and Georgia: Geographic Setting, 1994

Location at the meeting point of southeastern Europe with the western border of Asia greatly influenced the histories of the three national groups forming the present-day Transcaucasian republics (see figure 1; figure 2). Especially between the twelfth and the twentieth centuries, their peoples were subject to invasion and control by the Ottoman, Persian, and Russian empires. But, with the formation of the twentieth-century states named for them, the Armenian, Azerbaijani, and Georgian peoples as a whole underwent different degrees of displacement and played quite different roles. For example, the Republic of Azerbaijan that emerged from the Soviet Union in 1991 contains only 5.8 million of the world's estimated 19 million Azerbaijanis, with most of the balance living in Iran across a southern border fixed when Persia and Russia in the nineteenth century. At the same time, slightly more than half the world's 6.3 million Armenians are widely scattered outside the borders of the Republic of Armenia as a result of a centuries-long diaspora and step-by-step reduction of their national territory. In contrast, the great majority of the world's Georgian population lives in the Republic of Georgia (together with ethnic minorities constituting about 30 percent of the republic's population), after having experienced centuries of foreign domination but little forcible alteration of national boundaries.

Figure 2. Armenia, Azerbaijan, and Georgia: Topography and Drainage

The starting points and the outside influences that formed the three cultures also were quite different. In pre-Christian times, Georgia's location along the Black Sea opened it to cultural influence from Greece. During the same period, Armenia was settled by tribes from southeastern Europe, and Azerbaijan was settled by Asiatic Medes, Persians, and Scythians. In Azerbaijan, Persian cultural influence dominated in the formative period of the first millennium B.C. In the early fourth century, kings of Armenia and Georgia accepted Christianity after extensive contact with the proselytizing early Christians at the eastern end of the Mediterranean. Following their conversion, Georgians remained tied by religion to the Roman Empire and later the Byzantine Empire centered at Constantinople. Although Armenian Christianity broke with Byzantine Orthodoxy very early, Byzantine occupation of Armenian territory enhanced the influence of Greek culture on Armenians in the Middle Ages.

In Azerbaijan, the Zoroastrian religion, a legacy of the early Persian influence there, was supplanted in the seventh century by the Muslim faith introduced by conquering Arabs. Conquest and occupation by the Turks added centuries of Turkic influence, which remains a primary element of secular Azerbaijani culture, notably in language and the arts. In the twentieth century, Islam remains the prevalent religion of Azerbaijan, with about three-quarters of the population adhering to the Shia (see Glossary) branch.

Golden ages of peace and independence enabled the three civilizations to individualize their forms of art and literature before 1300, and all have retained unique characteristics that arose during those eras. The Armenian, Azerbaijani, and Georgian languages also grew in different directions: Armenian developed from a combination of Indo-European and non-Indo-European language stock, with an alphabet based on the Greek; Azerbaijani, akin to Turkish and originating in Central Asia, now uses the Roman alphabet after periods of official usage of the Arabic and Cyrillic alphabets; and Georgian, unrelated to any major world language, use a Greek-based alphabet quite different from the Armenian.

Beginning in the eighteenth century, the Russian Empire constantly probed the Caucasus region for possible expansion toward the Black Sea and the Caspian Sea. These efforts engaged Russia in a series of wars with the Persian and Ottoman empires, both of which by that time were decaying from within. By 1828 Russia had annexed or had been awarded by treaty all of present- day Azerbaijan and Georgia and most of present-day Armenia. (At that time, much of the Armenian population remained across the border in the Ottoman Empire.)

Except for about two years of unstable independence following World War I, the Transcaucasus countries remained under Russian, and later Soviet, control until 1991. As part of the Soviet Union from 1922 to 1991, they underwent approximately the same degree of economic and political regimentation as the other constituent republics of the union (until 1936 the Transcaucasian Soviet Federated Socialist Republic included all three countries). The Sovietization process included intensive industrialization, collectivization of agriculture, and large-scale shifts of the rural work force to industrial centers, as well as expanded and standardized systems for education, health care, and social welfare. Although industries came under uniform state direction, private farms in the three republics, especially in Georgia, remained important agriculturally because of the inefficiency of collective farms.

The achievement of independence in 1991 left the three republics with inefficient and often crumbling remains of the Soviet-era state systems. In the years that followed, political, military, and financial chaos prevented reforms from being implemented in most areas. Land redistribution proceeded rapidly in Armenia and Georgia, although agricultural inputs often remained under state control. In contrast, in 1994 Azerbaijan still depended mainly on collective farms. Education and health institutions remained substantially the same centralized suppliers as they had in the Soviet era, but availability of educational and medical materials and personnel dropped sharply after 1991. The military conflict in Azerbaijan's Nagorno- Karabakh Autonomous Region put enormous stress on the health and social welfare systems of combatants Armenia and Azerbaijan, and Azerbaijan's blockade of Armenia, which began in 1989, caused acute shortages of all types of materials (see figure 3).

Figure 3. Nagorno-Karabakh, 1994

The relationship of Russia to the former Soviet republics in the Transcaucasus caused increasing international concern in the transition years. The presence of Russian peacekeeping troops between Georgian and Abkhazian separatist forces remained an irritation to Georgian nationalists and an indication that Russia intended to intervene in that part of the world when opportunities arose. Russian nationalists saw such intervention as an opportunity to recapture nearby parts of the old Russian, and later Soviet empire. In the fall of 1994, in spite of strong nationalist resistance in each of the Transcaucasus countries, Russia was poised to improve its economic and military influence in Armenia and Azerbaijan, as it had in Georgia, if its mediation activities in Nagorno-Karabakh bore fruit.

The countries of Transcaucasia each inherited large state- owned enterprises specializing in products assigned by the Soviet system: military electronics and chemicals in Armenia, petroleum- based and textile industries in Azerbaijan, and chemicals, machine tools, and metallurgy in Georgia. As in most of the nations in the former Soviet sphere, redistribution and revitalization of such enterprises proved a formidable obstacle to economic growth and foreign investment in Armenia, Azerbaijan, and Georgia. Efforts at enterprise privatization were hindered by the stresses of prolonged military engagements, the staying power of underground economies that had defied control under communist and governments, the lack of commercial expertise, and the lack of a legal infrastructure on which to base new business relationships. As a result, in 1994 the governments were left with oversized, inefficient, and often bankrupt heavy industries whose operation was vital to provide jobs and to revive the national economies. At the same time, small private enterprises were growing rapidly, especially in Armenia and Georgia.

In the early 1990s, the Caucasus took its place among the regions of the world having violent post-Cold War ethnic conflict. Several wars broke out in the region once Soviet authority ceased holding the lid on disagreements that had been fermenting for decades. (Joseph V. Stalin's forcible relocation of ethnic groups after the redrawing of the region's political map was a chief source of the friction of the 1990s.) Thus, the three republics devoted critical resources to military campaigns in a period when the need for internal restructuring was paramount.

In Georgia, minority separatist movements--primarily on the part of the Ossetians and the Abkhaz, both given intermittent encouragement by the Soviet regime over the years--demanded fuller recognition in the new order of the early 1990s. Asserting its newly gained national prerogatives, Georgia responded with military attempts to restrain separatism forcibly. A year-long battle in South Ossetia, initiated by Zviad Gamsakhurdia, post- Soviet Georgia's ultranationalist first president, reached an uneasy peace in mid-1992. Early in 1992, however, the violent eviction of Gamsakhurdia from the presidency added another opponent of Georgian unity as the exiled Gamsakhurdia gathered his forces across the border.

In mid-1992 Georgian paramilitary troops entered the Abkhazian Autonomous Republic of Georgia, beginning a new conflict that in 1993 threatened to break apart the country. When Georgian troops were driven from Abkhazia in September 1993, Georgia's President Eduard Shevardnadze was able to gain Russian military aid to prevent the collapse of the country. In mid-1994 an uneasy cease-fire was in force; Abkhazian forces controlled their entire region, but no negotiated settlement had been reached. Life in Georgia had stabilized, but no permanent answers had been found to ethnic claims and counterclaims.

For Armenia and Azerbaijan, the center of nationalist self- expression in this period was the Nagorno-Karabakh Autonomous Region of Azerbaijan. After the Armenian majority there declared unification with Armenia in 1988, ethnic conflict broke out in both republics, leaving many Armenians and Azerbaijanis dead. For the next six years, battles raged between Armenian and Azerbaijani regular forces and between Armenian militias from Nagorno-Karabakh ("mountainous Karabakh" in Russian), and foreign mercenaries, killing thousands in and around Karabakh and causing massive refugee movements in both directions. Armenian military forces, better supplied and better organized, generally gained ground in the conflict, but the sides were evened as Armenia itself was devastated by six years of Azerbaijani blockades. In 1993 and early 1994, international mediation efforts were stymied by the intransigence of the two sides and by competition between Russia and the Conference on Security and Cooperation in Europe (CSCE--see Glossary) for the role of chief peace negotiator.

Armenia

Armenia, in the twentieth century the smallest of the three republics in size and population, has undergone the greatest change in the location of its indigenous population. After occupying eastern Anatolia (now eastern Turkey) for nearly 2,000 years, the Armenian population of the Ottoman Empire was extinguished or driven out by 1915 adding to a diaspora that had begun centuries earlier. After 1915, only the eastern population, in and around Erevan, remained in its original location. In the Soviet era, Armenians preserved their cultural traditions, both in Armenia and abroad. The Armenian people's strong sense of unity has been reinforced by periodic threats to their existence. When Armenia, Azerbaijan, and Georgia gained their independence in 1991, Armenia possessed the fewest natural and man-made resources upon which to build a new state. Fertile agricultural areas are relatively small, transportation is limited by the country's landlocked position and mountainous terrain (and, beginning in 1989, by the Azerbaijani blockade), and the material base for industry is not broad. A high percentage of cropland requires irrigation, and disorganized land privatization has delayed the benefits that should result from reducing state agricultural control. Although harvests were bountiful in 1993, gaps in support systems for transport and food processing prevented urban populations from benefiting.

The intensive industrialization of Armenia between the world wars was accomplished within the controlled barter system of the Soviet republics, not within a separate economic unit. The specialized industrial roles assigned Armenia in the Soviet system offered little of value to the world markets from which the republic had been protected until 1991. Since 1991 Armenia has sought to reorient its Soviet-era scientific-research, military electronics, and chemicals infrastructures to satisfy new demands, and international financial assistance has been forthcoming. In the meantime, basic items of Armenian manufacturing, such as textiles, shoes, and carpets, have remained exportable. However, the extreme paucity of energy sources--little coal, natural gas, or petroleum is extracted in Armenia--always has been a severe limitation to industry. And about 30 percent of the existing industrial infrastructure was lost in the earthquake of 1988. Desperate crises arose throughout society when Azerbaijan strangled energy imports that had provided over 90 percent of Armenia's energy. Every winter of the early 1990s brought more difficult conditions, especially for urban Armenians.

In the early 1990s, the Armenian economy was also stressed by direct support of Karabakh self-determination Karabakh, which received massive shipments of food and other materials through the Lachin corridor that Karabakh Armenian forces had opened across southwestern Azerbaijan. Although Karabakh sent electricity to Armenia in return, the balance of trade was over two to one in favor of Karabakh, and Armenian credits covered most of Karabakh's budget deficits. Meanwhile, Armenia remained a command rather than a free-market economy to ensure that the military received adequate economic support.

In addition to the Karabakh conflict, wage, price, and social welfare conditions have caused substantial social unrest since independence. The dram (for value of the dram--see Glossary), the national currency introduced in 1992, underwent almost immediate devaluation as the national banking system tried to stabilize international exchange rates. Accordingly, in 1993 prices rose to an average of 130 percent of wages, which the government indexed through that year. The scarcity of many commodities, caused by the blockade, also pushed prices higher. In the first post-Soviet years, and especially in 1993, plant closings and the energy crisis caused unemployment to more than double. At the same time, the standard of living of the average Armenian deteriorated; by 1993 an estimated 90 percent of the population were living below the official poverty line.

Armenia's first steps toward democracy were uneven. Upon declaring independence, Armenia adapted the political system, set forth in its Soviet-style 1978 constitution, to the short- term requirements of governance. The chief executive would be the chairman of Armenia's Supreme Soviet, which was the chief legislative body of the new republic--but in independent Armenia the legislature and the executive branch would no longer merely rubber-stamp policy decisions handed down from Moscow.

The inherited Soviet system was used in the expectation that a new constitution would prescribe Western-style institutions in the near future. However, between 1992 and 1994 consensus was not reached between factions backing a strong executive and those backing a strong legislature.

At the center of the dispute over the constitution was Levon Ter-Petrosian, president (through late 1994) of post-Soviet Armenia. Beginning in 1991, Ter-Petrosian responded to the twin threats of political chaos and military defeat at the hands of Azerbaijan by accumulating extraordinary executive powers. His chief opposition, a faction that was radically nationalist but held few seats in the fragmented Supreme Soviet, sought to build coalitions to cut the president's power, then to finalize such a move in a constitution calling for a strong legislature. As they had on other legislation, however, the chaotic deliberations of parliament yielded no decision. Ter-Petrosian was able to continue his pragmatic approach to domestic policy, privatizing the economy whenever possible, and to continue his moderate, sometimes conciliatory, tone on the Karabakh issue.

Beginning in 1991, Armenia's foreign policy also was dictated by the Karabakh conflict. After independence, Russian troops continued serving as border guards and in other capacities that Armenia's new national army could not fill. Armenia, a charter member of the Russian-sponsored Commonwealth of Independent States (CIS--see Glossary), forged security agreements with CIS member states and took an active part in the organization. After 1991 Russia remained Armenia's foremost trading partner, supplying the country with fuel. As the Karabakh conflict evolved, Armenia took a more favorable position toward Russian leadership of peace negotiations than did Azerbaijan.

The dissolution of the Soviet Union made possible closer relations with Armenia's traditional enemy Turkey, whose membership in the North Atlantic Treaty Organization (NATO--see Glossary) had put it on the opposite side in the Cold war. In the Karabakh conflict, Turkey sided with Islamic Azerbaijan, blocking pipeline deliveries to Armenia through its territory. Most important,

Turkey withheld acknowledgment of the 1915 massacre, without which no Armenian government could permit a rapprochement. Nevertheless, tentative contacts continued throughout the early 1990s.

In spite of pressure from nationalist factions, the Ter- Petrosian government held that Armenia should not unilaterally annex Karabakh and that the citizens of Karabakh had a right to self-determination (presumably meaning either independence or union with Armenia). Although Ter-Petrosian maintained contact with Azerbaijan's President Heydar Aliyev, and Armenia officially accepted the terms of several peace proposals, recriminations for the failure of peace talks flew from both sides in 1993.

The United States and the countries of the European Union (EU) have aided independent Armenia in several ways, although the West has criticized Armenian incursions into Azerbaijani territory. Humanitarian aid, most of it from the United States, played a large role between 1991 and 1994 in Armenia's survival through the winters of the blockade. Armenia successively pursued aid from the European Bank for Reconstruction and Development, the International Monetary Fund (IMF--see Glossary), and the World Bank (see Glossary). Two categories of assistance, humanitarian and technical, were offered through those lenders. Included was aid for recovery from the 1988 earthquake, whose destructive effects were still being felt in Armenia's industry and transportation infrastructure as of late 1994.

After the Soviet Union collapsed, Armenia's national security continued to depend heavily on the Russian military. The officer corps of the new national army created in 1992 included many Armenian former officers of the Soviet army, and Russian institutes trained new Armenian officers. Two Russian divisions were transferred to Armenian control, but another division remained under full Russian control on Armenian soil.

Internal security was problematic in the transitional years. The Ministry of Internal Affairs, responsible for internal security agencies, remained outside regular government control, as it had been in the Soviet period. This arrangement led to corruption, abuses of power, and public cynicism, a state of affairs that was especially serious because the main internal security agency acted as the nation's regular police force. The distraction of the Karabakh crisis combined with security lapses to stimulate a rapid rise in crime in the early 1990s. The political situation was also complicated by charges of abuse of power exchanged by high government officials in relation to security problems.

By the spring of 1994, Armenians had survived a fourth winter of acute shortages, and Armenian forces in Karabakh had survived the large-scale winter offensive that Azerbaijan launched in December 1993. In May 1994, a flurry of diplomatic activity by Russia and the CIS, stimulated by the new round of fighting, produced a cease-fire that held, with some violations, through the summer. A lasting treaty was delayed, however, by persistent disagreement over the nationality of peacekeeping forces that would occupy Azerbaijan. Azerbaijan resisted the return of Russian troops to its territory, while the Russian plan called for at least half the forces to be Russian. On both diplomatic and economic fronts, new signs of stability caused guarded optimism in Armenia in the fall of 1994.

The failure of the CSCE peace plan, which Azerbaijan supported, had caused that country to mount an all-out, human- wave offensive in December 1993 and January 1994, which initially pushed back Armenian defensive lines in Karabakh and regained some lost territory. When the offensive stalled in February, Russia's minister of defense, Pavel Grachev, negotiated a cease- fire, which enabled Russia to supplant the CSCE as the primary peace negotiator. Intensive Russian-sponsored talks continued through the spring, although Azerbaijan mounted air strikes on Karabakh as late as April. In May 1994, Armenia, Azerbaijan, and Nagorno-Karabakh signed the CIS-sponsored Bishkek Protocol, calling for a cease-fire and the beginning of troop withdrawals. In July the defense ministers of the three jurisdictions officially extended the cease-fire, signaling that all parties were moving toward some combination of the Russian and the CSCE peace plans. In September the exchange of Armenian and Azerbaijani prisoners of war began.

Under these conditions, Russia was able to intensify its three-way diplomatic gambit in the Transcaucasus, steadily erasing Armenians' memory of airborne Soviet forces landing unannounced as a show of strength in 1991. In the first half of 1994, Armenia moved closer to Russia on several fronts. A February treaty established bilateral barter of vital resources. In March Russia agreed to joint operation of the Armenian Atomic Power Station at Metsamor, whose scheduled 1995 reopening is a vital element in easing the country's energy crisis. Also in March, Armenia replaced its mission in Moscow with a full embassy. In June the Armenian parliament approved the addition of airborne troops to the Russian garrison at Gyumri near the Turkish border. Then in July, Russia extended 100 billion rubles (about US$35 million at that time) for reactivation of the Metsamor station, and Armenia signed a US$250 million contract with Russia for Armenia to process precious metals and gems supplied by Russia. In addition, Armenia consistently favored the Russian peace plan for Nagorno-Karabakh, in opposition to Azerbaijan's insistence on reviving the CSCE plan that prescribed international monitors rather than combat troops (most of whom would be Russian) on Azerbaijani soil.

Armenia was active on other diplomatic fronts as well in 1994. President Ter-Petrosian made official visits to Britain's Prime Minister John Major in February (preceding Azerbaijan's Heydar Aliyev by a few weeks when the outcome of the last large- scale campaign in the Karabakh conflict remained in doubt) and to President William J. Clinton in the United States in August. Clinton promised more active United States support for peace negotiations, and an exchange of military attachés was set. While in Washington, Ter-Petrosian expressed interest in joining the NATO Partnership for Peace, in which Azerbaijan had gained membership three months earlier.

Relations with Turkey remained cool, however. In 1994 Turkey continued its blockade of Armenia in support of Azerbaijan and accused Armenia of fostering rebel activity by Kurdish groups in eastern Turkey; it reiterated its denial of responsibility for the 1915 massacre of Armenians in the Ottoman Empire. In June these policies prompted Armenia to approve the security agreement with Russia that stationed Russian airborne troops in Armenia near the Turkish border. In July Armenia firmly refused Turkey's offer to send peacekeeping forces to Nagorno-Karabakh. Thus, Armenia became an important player in the continuing contest between Russia and Turkey for influence in the Black Sea and Caucasus regions. Armenians considered the official commemoration by Israel and Russia of the 1915 Armenian massacre a significant advancement in the country's international position.

Early in 1994, Armenia's relations with Georgia worsened after Azerbaijani terrorists in Georgia again sabotaged the natural gas pipeline supplying Armenia through Georgia. Delayed rail delivery to Armenia of goods arriving in Georgian ports also caused friction. Underlying these stresses were Georgia's unreliable transport system and its failure to prevent violent acts on Georgian territory. Pipeline and railroad sabotage incidents continued through mid-1994.

The domestic political front remained heated in 1994. As the parliamentary elections of 1995 approached, Ter-Petrosian's centrist Armenian Pannational Movement (APM), which dominated political life after 1991, had lost ground to the right and the left because Armenians were losing patience with economic hardship. Opposition newspapers and citizens' groups, which Ter- Petrosian refused to outlaw, continued their accusations of official corruption and their calls for the resignation of the Ter-Petrosian government early in the year. Then, in mid-1994 the opposition accelerated its activity by mounting antigovernment street demonstrations of up to 50,000 protesters.

In the protracted struggle over a new constitution, the opposition intensified rhetoric supporting a document built around a strong legislature rather than the strong-executive version supported by Ter-Petrosian. By the fall of 1994, little progress had been made even on the method of deciding this critical issue. While opposition parties called for a constitutional assembly, the president offered to hold a national referendum, following which he would resign if defeated.

Economic conditions were also a primary issue for the opposition. The value of the dram, pegged at 14.5 to the United States dollar when it was established in November 1993, had plummeted to 390 to the dollar by May 1994. In September a major overhaul of Armenia's financial system was under way, aimed at establishing official interest rates and a national credit system, controlling inflation, opening a securities market, regulating currency exchange, and licensing lending institutions. In the overall plan, the Central Bank of Armenia and the Erevan Stock Exchange assumed central roles in redirecting the flow of resources toward production of consumer goods. And government budgeting began diverting funds from military to civilian production support, a step advertised as the beginning of the transition from a command to a market economy. This process included the resumption of privatization of state enterprises, which had ceased in mid-1992, including full privatization of small businesses and cautious partial privatization of larger ones. In mid-1994 the value of the dram stabilized, and industrial production increased somewhat. As another winter approached, however, the amount of goods and food available to the average consumer remained at or below subsistence level, and social unrest threatened to increase.

In September Armenia negotiated terms for the resumption of natural gas deliveries from its chief supplier, Turkmenistan, which had threatened a complete cutoff because of outstanding debts. Under the current agreement, all purchases of Turkmen gas were destined for electric power generation in Armenia. Also in September, the IMF offered favorable interest rates on a loan of US$800 million if Armenia raised consumer taxes and removed controls on bread prices. Armenian officials resisted those conditions because they would further erode living conditions.

Thus in mid-1994 Armenia, blessed with strong leadership and support from abroad but cursed with a poor geopolitical position and few natural resources, was desperate for peace after the Karabakh Armenians had virtually won their war for self- determination. With many elements of post-Soviet economic reform in place, a steady flow of assistance from the West, and an end to the Karabakh conflict in sight, Armenia looked forward to a new era of development.

Azerbaijan

Azerbaijan, the easternmost and largest of the Transcaucasus states in size and in population, has the richest combination of agricultural and industrial resources of the three states. But Azerbaijan's quest for reform has been hindered by the limited contact it had with Western institutions and cultures before the Soviet era began in 1922.

Although Azerbaijan normally is included in the three-part grouping of the Transcaucasus countries (and was so defined politically between 1922 and 1936), it has more in common culturally with the Central Asian republics east of the Caspian Sea than with Armenia and Georgia. The common link with the latter states is the Caucasus mountain range, which defines the topography of the northern and western parts of Azerbaijan. A unique aspect of Azerbaijan's political geography is the enclave of the Nakhichevan Autonomous Republic, created by the Soviet Union in 1924 in the area between Armenia and Iran and separated from the rest of Azerbaijan by Armenian territory. In 1924 the Soviet Union also created the Nagorno-Karabakh Autonomous Region within Azerbaijan, an enclave whose population was about 94 percent Armenian at that time and remained about 75 percent Armenian in the late 1980s.

Beginning in the last years of the Soviet Union and extending into the 1990s, the drive for independence by Nagorno-Karabakh's Armenian majority was an issue of conflict between Armenia, which insisted on self-determination for its fellow Armenians, and Azerbaijan, which cited historical acceptance of its sovereignty whatever the region's ethnic composition. By the 1991 independence struggle was an issue of de facto war between Azerbaijan and the Karabakh Armenians, who by 1993 controlled all of Karabakh and much of adjoining Azerbaijan.

The population of Azerbaijan, already 83 percent Azerbaijani before independence, became even more homogeneous as members of the two principal minorities, Armenians and Russians, emigrated in the early 1990s and as thousands of Azerbaijanis immigrated from neighboring Armenia. The heavily urbanized population of Azerbaijan is concentrated around the cities of Baku, Gyandzha, and Sumgait.

Like the other former Soviet republics, Azerbaijan began in 1991 to seek the right combination of indigenous and "borrowed" qualities to replace the awkwardly imposed economic and political imprint of the Soviet era. And, like Armenia and Georgia, Azerbaijan faced the complications of internal political disruption and military crisis in the first years of this process.

For more than 100 years, Azerbaijan's economy has been dominated by petroleum extraction and processing. In the Soviet system, Azerbaijan's delegated role had evolved from supplying crude oil to supplying oil-extraction equipment, as Siberian oil fields came to dominate the Soviet market and as Caspian oil fields were allowed to deteriorate. Although exploited oil deposits were greatly depleted in the Soviet period, the economy still depends heavily on industries linked to oil. The country also depends heavily on trade with Russia and other former Soviet republics. Azerbaijan's overall industrial production dropped in the early 1990s, although not as drastically as that of Armenia and Georgia. The end of Soviet-supported trade connections and the closing of inefficient factories caused unemployment to rise and industrial productivity to fall an estimated 26 percent in 1992; acute inflation caused a major economic crisis in 1993.

Azerbaijan did not restructure its agriculture as quickly as did Armenia and Georgia; inefficient Soviet methods continued to hamper production, and the role of private initiative remained small. Agriculture in Azerbaijan also was hampered by the conflict in Nagorno-Karabakh, which was an important source of fruits, grain, grapes, and livestock. As much as 70 percent of Azerbaijan's arable land was occupied by military forces at some stage of the conflict.

In spite of these setbacks, Azerbaijan's economy remains the healthiest among the three republics, largely because unexploited oil and natural gas deposits are plentiful (although output declined in the early 1990s) and because ample electric-power generating plants are in operation. Azerbaijan has been able to attract Western investment in its oil industry in the post-Soviet years, although Russia remains a key oil customer and investor. In 1993 the former Soviet republics remained Azerbaijan's most important trading partners, and state bureaucracies still controlled most foreign trade. Political instability in Baku, however, continued to discourage Turkey, a natural trading partner, from expanding commercial relations.

The political situation of Azerbaijan was extremely volatile in the first years of independence. With performance in Nagorno- Karabakh rather than achievement of economic and political reform as their chief criterion, Azerbaijanis deposed presidents in 1992 and 1993, then returned former communist party boss Heydar Aliyev to power. In 1992, in the country's first and only free election, the people had chosen Abulfaz Elchibey, leader of the Azerbaijani Popular Front (APF), as president. Meanwhile, the Azerbaijani Communist Party, formally disbanded in 1991, retained positions of political and economic power and was key in the coup that returned Aliyev to power in June 1993. Former communists dominated policy making in the government Aliyev formed after his rubber-stamp election as president the following October. However, the APF remained a formidable opposition force, especially critical of any sign of weakness on the Nagorno- Karabakh issue.

During the transition period, the only national legislative body was the Melli-Majlis (National Council), a fifty-member interim assembly that came under the domination of former communists and, by virtue of postponing parliamentary elections indefinitely, continued to retain its power in late 1994. Aliyev promised a new constitution and democratic rule, but he prolonged his dictatorial powers on the pretext of the continuing military emergency. Work on a new constitution was begun in 1992, but the Nagorno-Karabakh conflict and political turmoil delayed its completion; meanwhile, elements of the 1978 constitution (based on the 1977 constitution of the Soviet Union) remain the highest law of the land, supplemented only by provisions of the 1991 Act of Independence.

Azerbaijan's post-Soviet foreign policy attempted to balance the interests of three stronger, often mutually hostile, neighbors--Iran, Russia, and Turkey--while using those nations' interests in regional peace to help resolve the Karabakh conflict. The Elchibey regime of 1992-93 leaned toward Turkey, which it saw as the best mediator in Karabakh. Armenia took advantage of this strategy, however, to form closer ties with Russia, whose economic assistance it needed desperately. Beginning in 1993, Aliyev sought to rekindle relations with Russia and Iran, believing that Russia could negotiate a positive settlement in Karabakh. Relations with Turkey were carefully maintained, however.

Beginning in 1991, Azerbaijan's external national security was breached by the incursion of the Armenian separatist forces of Karabakh militias and reinforcements from Armenia. Azerbaijan's main strategy in this early period was to blockade landlocked Armenia's supply lines and to rely for national defense on the Russian 4th Army, which remained in Azerbaijan in 1991. Clashes between Russian troops and Azerbaijani civilians in 1991 and the collapse of the Soviet Union, however, led Russia to a rapid commitment for withdrawal of troops and equipment, which was completed in mid-1993.

Under those circumstances, a new, limited national armed force was planned in 1992, and, as had been done in Armenia, the government appealed to Azerbaijani veterans of the Soviet army to defend their homeland. But the force took shape slowly, and outside assistance--mercenaries and foreign training officers-- were summoned to stem the Armenian advance that threatened all of southern Azerbaijan. In 1993 continued military failures brought reports of mass desertion and subsequent large-scale recruitment of teenage boys, as well as wholesale changes in the national defense establishment.

In the early 1990s, the domestic and international confusion bred by the Karabakh conflict increased customs violations, white-collar crime, and threats to the populace by criminal bands. The role of Azerbaijanis in the international drug market expanded noticeably. In 1993 the Aliyev government responded to these problems with a major reform of the Ministry of Internal Affairs, which had been plagued by corruption and incompetence, but experts agreed that positive results required a more stable overall atmosphere.

In December 1993, Azerbaijan launched a major surprise attack on all fronts in Karabakh, using newly drafted personnel in wave attacks, with air support. The attack initially overwhelmed Armenian positions in the north and south but ultimately was unsuccessful. An estimated 8,000 Azerbaijani troops died in the two-month campaign, which Armenian authorities described as Azerbaijan's best-planned offensive of the conflict.

When the winter offensive failed, Aliyev began using diplomatic channels to seek peace terms acceptable to his constituents, involving Russia as little as possible. Already in March, the chairman of the Azerbaijani parliament had initiated a private meeting with his opposite number from Armenia, an event hailed in the Azerbaijani press as a major Azerbaijani peace initiative. Official visits by Aliyev to Ankara and London early in 1994 yielded little additional support for Azerbaijan's position. (Turkey remained suspicious of Aliyev's communist background.)

At this point, Azerbaijan reasserted its support for the CSCE peace plan, which would use international monitors rather than military forces to enforce the cease-fire in Karabakh. Perhaps with the goal of avoiding further military losses, Aliyev approved in May the provisional cease-fire conditions of the Bishkek Protocol, sponsored by the CIS. That agreement, which softened Azerbaijan's position on recognizing the sovereignty of Nagorno-Karabakh, was subsequently the basis for terms of a true armistice.

Azerbaijan's official position on armistice conditions remained unchanged, however, during the negotiations of the summer and fall of 1994, in the face of Armenia's insistence that only an armed peacekeeping force (inevitably Russian) could prevent new outbreaks of fighting. During that period, sporadic Azerbaijani attacks tended to confirm Armenia's judgment. At the same time, Aliyev urged that his countrymen take a more conciliatory position toward Russia. Aliyev argued that the Soviet Union, not Russia, had sent the troops who had killed Azerbaijanis when they arrived to keep peace with Armenia in 1990 and that Azerbaijan could profit from exploiting rather than rejecting the remaining ties between the two countries.

In May Aliyev signed the NATO Partnership for Peace agreement, giving Azerbaijan the associate status that NATO had offered to East European nations and the former republics of the Soviet Union in late 1993. The same month, Aliyev received a mid- level United States delegation charged with discussing diplomatic support for the Nagorno-Karabakh peace process, Caspian Sea oil exploration by United States firms, and bilateral trade agreements.

In July Aliyev extended his diplomacy to the Muslim world, visiting Saudi Arabia and Iran in an effort to balance his diplomatic contacts with the West. Iran was especially important because of its proximity to Karabakh and its interest in ending the conflict on its border. Iran responded to offers of economic cooperation by insisting that any agreement must await a peace treaty between Azerbaijan and Armenia.

In the fall of 1994, a seventeen-point peace agreement was drafted, but major issues remained unresolved. Azerbaijani concerns centered on withdrawal of Armenian forces from Azerbaijani territory and conditions that would permit Azerbaijani refugees to return home. (An estimated 1 million Azerbaijanis had fled to other parts of Azerbaijan or Iran from occupied territory.) The top priorities for Armenia were ensuring security for Armenians in Karabakh and defining the status of the region prior to the withdrawal of forces.

A second result of the failed winter offensive of 1993-94 was a new crackdown by the Aliyev government on dissident activity. Early in 1994, censors in the Main Administration for Protecting State Secrets in the Press sharply increased censorship of material criticizing the regime, and the government cut the supply of paper and printing plates to opposition newspapers. In May a confrontation between Aliyev loyalists and opponents in the Melli-Majlis resulted in arrests of opposition leaders and reduction in the number of members required for a quorum to pass presidential proposals.

The issue behind the May dispute was Aliyev's handling of the Karabakh peace process. A variety of opposition parties and organizations claimed that the Bishkek Protocol had betrayed Azerbaijan by recognizing the sovereignty of Nagorno-Karabakh. A new coalition, the National Resistance Movement, was formed immediately after the May confrontation in the Mellis-Majlis. The movement's two principles were opposition to reintroduction of Russian forces in Azerbaijan and opposition to Aliyev's "dictatorship." By the end of the summer, however, the movement had drawn closer to Aliyev's position on the first point, and the announcement of long-delayed parliamentary elections to be held in the summer of 1995 aimed to defuse charges of dictatorship. Draft election legislation called for replacing the "temporary" Melli-Majlis with a 150-seat legislature in 1995.

In October 1994, a military coup, supported by Prime Minister Suret Huseynov, failed to topple Aliyev. Aliyev responded by declaring a two-month state of emergency, banning demonstrations, and taking military control of key positions. Huseynov, who had signed the Bishkek Protocol as Azerbaijan's representative, was dismissed.

Price and wage levels continued to reduce the standard of living in Azerbaijan in 1994. Between mid-1993 and mid-1994, prices increased by an average of about sixteen times; from November 1993 to July 1994, the state-established minimum wage more than doubled. To speed conversion to a market economy, the ministries of finance and economics submitted plans in July to combine state-run enterprises in forms more suitable for privatization. Land privatization has proceeded cautiously because of strong political support for maintaining the Soviet- era state-farm system. In mid-1994 about 1 percent of arable land was in private hands, the bureaucratic process for obtaining private land remained long and cumbersome, and state allocation of equipment to private farmers was meager.

Meanwhile, in 1994 currency-exchange activity increased dramatically in Azerbaijani banks, bringing more foreign currency into the country. The ruble remained the most widely used foreign unit in 1994. In June, at the insistence of the IMF and the World Bank, the National Bank of Azerbaijan stopped issuing credit that lacked monetary backing, a practice that had fueled inflation and destabilized the economy.

The main hope for Azerbaijan's economic recovery lies in reviving exploitation of offshore oil deposits in the Caspian Sea. By 1993 these deposits had attracted strong interest among British, Norwegian, Russian, Turkish, and United States firms. Within a consortium of such firms, Russia would likely have a 10 percent share and provide the pipeline and the main port (Novorossiysk on the Black Sea) for export of Azerbaijan's oil. An agreement signed in September 1994 included United, British, Turkish, Russian, and Azerbaijani oil companies.

In the early 1990s, the development of Azerbaijan's foreign trade was skewed by the refusal of eighteen nations, including the United States, Canada, Israel, India, and the Republic of Korea (South Korea), to import products from Azerbaijan as long as the blockade of Armenia continued. At the same time, many of those countries sold significant amounts of goods in Azerbaijan. Overall, in the first half of 1994 one-third of Azerbaijan's imports came from the "far abroad" (all non-CIS trading partners), and 46 percent of its exports went outside the CIS. In that period, total imports exceeded total exports by US$140 million. At the same time, the strongest long-term commercial ties within the CIS were with Kazakhstan, Russia, Turkmenistan, and Ukraine.

Like Armenia, Azerbaijan was able to improve internal conditions only marginally while awaiting the relief of a final peace settlement in Karabakh. Unlike either of its Transcaucasus neighbors, however, Azerbaijan had the prospect of major large- scale Western investment once investment conditions improved. Combined with potential oil earnings, diplomatic approaches by President Aliyev in 1994 to a number of foreign countries, including all of Azerbaijan's neighbors, seemed to offer it a much-improved postwar international position. A great deal depended, however, on the smooth surrender of wartime emergency powers by the Aliyev government and on accelerating the stalled development of a market economy.

Georgia

Georgia possesses the advantages of a subtropical Black Sea coastline and a rich mixture of Western and Eastern cultural elements. A combination of topographical and national idiosyncracies has preserved that cultural blend, whose chief impetus was the Georgian golden age of the twelfth and early thirteenth centuries, during long periods of occupation by foreign empires. Perhaps the most vivid result of this cultural independence is the Georgian language, unrelated to any other major tongue and largely unaffected by the languages of conquering peoples--at least until the massive influx of technical loanwords at the end of the twentieth century.

Since independence, Georgia has had difficulty establishing solid political institutions. This difficulty has been caused by the distractions of continuing military crises and by the chronic indecision of policy makers about the country's proper long-term goals and the strategy to reach them. Also, like the other Transcaucasus states, Georgia lacks experience with the democratic institutions that are now its political ideal; rubber- stamp passage of Moscow's agenda is quite different from creation of a legislative program useful to an emerging nation.

As in Azerbaijan, Georgia's most pressing problem has been ethnic separatism within the country's borders. Despite Georgia's modest size, throughout history all manifestations of a Georgian nation have included ethnic minorities that have conflicted with, or simply ignored, central power. Even in the golden age, when a central ruling power commanded the most widespread loyalty, King David the Builder was called "King of the Abkhaz, the Kartvelians, the Ran, the Kakhetians, and the Armenians." In the twentieth century, arbitrary rearrangement of ethnic boundaries by the Soviet regime resulted in the sharpening of various nationalist claims after Soviet power finally disappeared. Thus, in 1991 the South Ossetians of Georgia demanded union with the Ossetians across the Russian border, and in 1992 the Abkhaz of Georgia demanded recognition as an independent nation, despite their minority status in the region of Georgia they inhabited.

As in Armenia and Azerbaijan, influential, intensely nationalist factions pushed hard for unqualified military success in the struggle for separatist territory. And, as in the other Transcaucasus nations, those factions were frustrated by military and geopolitical reality: in Georgia's case, an ineffective Georgian army required assistance from Russia, the imperialist neighbor against whom nationalists had sharpened their teeth only three years earlier, to save the nation from fragmentation. At the end of 1993, Russia seemingly had settled into a long-term role of peacekeeping and occupation between Georgian and Abkhazian forces.

The most unsettling internal crisis was the failed presidency of Zviad Gamsakhurdia, once a respected human rights advocate and the undisputed leader of Georgia's nationalist opposition as the collapse of the Soviet Union became imminent. In 1991 Gamsakhurdia's dictatorial and paranoid regime, followed by the bloody process of unseating him, gave Georgia a lasting reputation for instability that damaged prospects for foreign investment and for participation in international organizations.

The failure of the one-year Gamsakhurdia regime necessitated a new political beginning that coincided with the establishment of Eduard Shevardnadze as head of state in early 1992. Easily the most popular politician in Georgia and facing chronically fragmented opposition in parliament, Shevardnadze acquired substantial "temporary" executive powers as he maneuvered to maintain national unity. At the same time, his hesitation to imitate Gamsakhurdia's grab for power often left a vacuum that was filled by quarreling splinter parties with widely varied agendas. Shevardnadze preserved parts of his reform program by forming temporary coalitions that dissolved when a contentious issue appeared. Despite numerous calls for his resignation, and despite rampant government corruption and frequent shifts in his cabinet between 1992 and 1994, there were no other serious contenders for Shevardnadze's position as of late 1994.

Shevardnadze also used familiarity with the world of diplomacy to reestablish international contacts, gain sympathy for Georgia's struggle to remain unified, and seek economic ties wherever they might be available. Unlike Armenia and Azerbaijan, Georgia did not arouse particular loyalty or hostility among any group of nations. In the first years of independence, Shevardnadze made special overtures to Russia, Turkey, and the United States and attempted to balance Georgia's approach to Armenia and Azerbaijan, its feuding neighbors in the Transcaucasus.

The collapse of the Soviet Union changed Georgia's economic position significantly, although industrial production already was declining in the last Soviet years. In the Soviet system, Georgia's assignment was mainly to supply the union with agricultural products, metal products, and the foreign currency collected by Georgian tourist attractions. This specialization made Georgia dependent on other Soviet republics for a wide range of products that were unavailable after 1991. Neither diversification nor meaningful privatization was possible, however, under the constant upheaval and energy shortages of the early 1990s. In addition, powerful organized criminal groups gained control of large segments of the national economy, including the export trade.

After the January 1992 fall of Gamsakhurdia's xenophobic regime, the maintenance of internal peace and unity was a critical national security issue. Although some progress was made in establishing a national armed force in 1994 the paramilitary organizations--the Mkhedrioni (horsemen) and the National Guard-- remained influential military forces in the fall of 1994. The small size and the poor organization of those groups had forced the request for Russian troop assistance in late 1993, which in turn renewed the national security dilemma of occupation by foreign troops. Meanwhile, civilian internal security forces, of which Shevardnadze took personal control in 1993, gained only partial victories over the crime wave that accompanied Georgia's post-Soviet upheavals. A series of reorganizations in security agencies failed to improve the protection of individuals against random crime or of the economic system against organized groups.

Through most of 1994, the Abkhazian conflict was more diplomatic than military. In spite of periodic hostilities, the uneasy truce line held along the Inguri River in far northwestern Georgia (in the campaign of October 1993, Georgian forces had been pushed out of all of Abkhazia except the far northern corner). The role of the 3,000 Russian peacekeepers on the border, and their relationship with United Nations (UN) observers, was recognized by a resolution of the UN Security Council in July. Throughout that period, the issue of the return of as many as 300,000 Georgian refugees to Abkhazia was the main sticking point of negotiations. The Abkhaz saw the influx of so many Georgians as a danger to their sovereignty, which Georgia did not recognize, and the refugees' plight as a bargaining chip to induce further Georgian withdrawal. No settlement was likely before the refugee issue was resolved. Meanwhile, supporting the refugees placed additional stress on Georgian society.

A legal basis for the presence of Russian troops in Georgia had been established in a status-of-forces treaty between the two nations in January 1994. The treaty prescribed the authority and operating conditions of the Group of Russian Troops in the Caucasus (GRTC), which was characterized as on Georgian territory for a "transitional period." In the summer of 1994, high-level bilateral talks covered Georgian-Russian military cooperation and further integration of CIS forces.

The Georgian economy continued to struggle in 1994, showing only isolated signs of progress. At the beginning of the year, state monopolies were reaffirmed in vital industries such as tea and food processing and electric power. By May, however, after prodding from the IMF, Shevardnadze began issuing decrees that eased privatization conditions. This policy spurred a noticeable acceleration of privatization in the summer of 1994. When the new stimulus began, about 23 percent of state enterprises had been privatized, and only thirty-nine joint-stock companies had formed out of the more than 900 large firms designated for that type of conversion. A voucher system for collecting private investment funds, delayed by a shortage of hard currency, finally began operating. But the state economic bureaucracy, entrenched since the Soviet era, was able to slow the privatization process when dispersal of economic power threatened its privileged position in 1994.

Between mid-1993 and mid-1994, prices rose by an average of 300 percent, and inflation severely eroded the government- guaranteed minimum wage. (In August the minimum wage, which was stipulated in coupons [for value of the coupon--see Glossary], equaled US$0.33 per month.) Often wages were withheld for months because of the currency shortage. In September the government raised price standards sharply for basic food items, transportation, fuel, and services. Lump-sum payments to all citizens, designed to offset this cost, failed to reach many, prompting new calls for Shevardnadze's resignation. Under those conditions, most Georgians were supported by a vast network of unofficial economic activities.

In mid-1994 unemployment was estimated unofficially at 1.5 million people, nearly 50 percent of Georgia's working-age population. The exchange rate of the Georgian coupon stabilized in early 1994 after many months of high inflation, but by that time the coupon had been virtually displaced in private transactions by the ruble and the dollar. The national financial system remained chaotic--especially in tax collection, customs, and import-export operations. The first major state bank was privatized in the summer of 1994. In August parliament approved a major reform program for social welfare, pricing, and the financial system.

In July 1994, a Georgian-Russian conference on economic cooperation discussed transnational corporations and concluded some contracts for joint economic activities, but most Russian investors demanded stronger legal guarantees for their risks. Numerous Western firms established small joint ventures in 1994, but the most critical investment project under discussion sought to exploit the substantial oil deposits that had been located by recent Australian, British, Georgian, and United States explorations in the Black Sea shelf near Batumi and Poti. A first step in foreign involvement, an oil refinery near Tbilisi, received funding in July, but the Western firms demanded major reform of commercial legislation before expanding their participation.

Georgia experienced a major energy crisis in the winter of 1993-94; following the crisis, in mid-1994 Turkmenistan drastically reduced natural gas supplies because of unpaid debts. Some fuel aid was expected for the winter of 1994-95 from Azerbaijan, the EU, Iran, and Turkey. The output of the domestic oil industry increased sharply in mid-1994. As winter approached, Georgia also offered Turkmenistan new assurances of payment in return for resumption of natural gas delivery.

Georgia's communications system, a chronically weak infrastructure link that also had discouraged foreign investment, began integration into world systems in early 1994 when the country joined international postal, satellite, and electronic communications organizations. Joint enterprises with Australian, French, German, Turkish, and United States communications companies allowed the upgrading of the national telephone system and installation of fiber-optic cables.

In the first half of 1994, the most frequent topic of government debate was the role of Russian troops in Abkhazia. By that time, opposition nationalist parties had accepted the Russian presence but rejected Abkhazian delays in allowing the return of refugees and Shevardnadze's tolerance of those delays. In May Shevardnadze overcame parliament's objections to new concessions to the Abkhaz by threatening to resign. The new agreement passed, and opposition leaders muted their demands for Shevardnadze's ouster in the belief that Russia was seeking to replace him with someone more favorable to Russian intervention. Nevertheless, in the fall of 1994 few Georgian refugees had returned to Abkhazia.

Shevardnadze's exercise of extraordinary executive powers remained a hot issue in parliament. One faction called for reduced powers in the name of democracy, but another claimed that a still stronger executive was needed to enforce order. In a July poll, 48 percent of respondents said the government was obstructing the mass media. Although the 1992 state of emergency continued to restrict dissemination of information, the Georgian media consistently presented various opposition views. Likewise, the Zviadists, Gamsakhurdia's supporters, although banned from radio and television, continued to hold rallies under the leadership of a young radical, Irakli Tsereteli.

In 1994 the government took steps to improve the internal security situation. In the latest of a long series of organizational and leadership shuffles, Shevardnadze replaced the Emergency Committee, which had been headed by former Mkhedrioni leader Jaba Ioseliani, with the Emergency Coordinating Commission, headed by Shevardnadze, and gave the commission a vague mandate to coordinate economic, political, defense, and law-enforcement matters. Ioseliani, whose command of the Mkhedrioni still gave him great influence, became a deputy head of the commission.

Shevardnadze's attempt to form a new, one-battalion Georgian army was delayed throughout the first half of 1994. The Ministry of Defense continued drafting potential soldiers (a very high percentage of whom evaded recruitment) for the Georgian armed forces and streamlining its organization. In September the national budget had not yet allocated wages, and sources of rations and equipment had not been identified--mainly because parliament had not passed the necessary legislation. Ministry of Defense plans called for the country's remaining state farms to be designated for direct military supply, as was the practice in the Soviet era. The disposition of existing paramilitary forces remained undecided as of late 1994.

The intelligence service had been reorganized in late 1993 to include elite troops mandated to fight drug smuggling and organized crime. In the spring of 1994, new agencies were formed in the State Security Service to investigate fiscal crimes and to combat terrorism. And in August 1994, the Ministry of Internal Affairs announced a major new drive against organized crime and drug traffickers throughout Georgia. Parliament and local jurisdictions offered indifferent support, however.

In 1994 Georgia began solving some of its most critical problems--laying a political base for a market economy, solidifying to a degree Shevardnadze's position as head of state, stabilizing inflation, and avoiding large-scale military conflict. But long-term stability will depend on comprehensive reform of the entire economy, eradication of the corruption that has pervaded both government and economic institutions, redirection of resources from the Abkhazian conflict into a civilian infrastructure suitable for international trade (and for major loans from international lenders), and, ultimately, finding political leaders besides Shevardnadze who are capable of focusing Georgians' attention on building a nation, rather than on advancing local interests. All those factors will influence the other major imponderable: Russia's long-term economic and political influence in Georgia, which increased greatly in late 1993 and in the first half of 1994. October 18, 1994

*   *   *

In the months following preparation of this manuscript, a number of significant events occurred in the three countries of the Transcaucasus. Cease-fires in two major conflicts, between Abkhazia and Georgia and between Armenia and Nagorno-Karabakh on one side and Azerbaijan on the other, remained in effect despite periodic hostilities. Although the two sets of peace talks continued to encounter fundamental differences, signs of compromise emerged from both in the first months of 1995, with the assistance of international mediators. All three countries continued efforts to stabilize their economies, reduce crime, and normalize political systems distorted by lengthy states of emergency.

At the beginning of 1995, Armenia had made the most progress toward economic recovery and political stability, although its population suffered another winter of privation because of Azerbaijan's fuel blockade. In December a summit of the Organisation for Security and Cooperation in Europe (OSCE, formerly the CSCE) had succeeded in merging OSCE and Russian peace efforts on Nagorno-Karabakh for the first time in an accord signed in Budapest. Russia was expected to become the head of the OSCE Minsk Group, which had been negotiating on behalf of Western Europe for the previous two years. In return, Russia accepted OSCE oversight of peacekeeping in the conflict zone. Armenia's President Ter-Petrosian reported the opening of three defense plants and full staffing of the Armenian Army in 1994, improving Armenia's national security position.

In November 1994, the World Bank announced loans to Armenia of US$265 million for infrastructural, agricultural, and energy applications. The bank cited Armenia's new reform program to control inflation and expand the private sector, together with the first increase in Armenia's gross national product (GNP--see Glossary) since independence, as the reasons for this investment. In December the reform package went into effect. Expected to improve the standing of President Ter-Petrosian's embattled government, the reform included substantial reduction of the government's budget deficit, which had caused many workers to go unpaid and others, including teachers, to accept barely subsistence wages. The second major reform measure was ending government subsidies for basic staples, including bread and utilities--a stringency measure highly unpopular in the short term but calculated to attract more international assistance. The price of bread rose by ten times as soon as the new law went into effect. In late 1994 and early 1995, Armenia also continued reestablishing commercial ties with Iran by signing a series of three economic treaties covering taxation, free trade, and capital investments. Beginning in 1992, commercial activity between the two countries had doubled annually, and the pace was expected to accelerate markedly in 1995.

Although the Armenian government had made more extensive preparations for another winter of hardship under the Azerbaijani blockade, conditions for the average Armenian were barely better than the year before. In the winter of 1994-95, Armenia's chronic fuel shortage, and the rising social unrest caused by it, were relieved somewhat by a new fuel agreement with Georgia and Turkmenistan. The pact provided for substantial increases in delivery of Turkmen natural gas through the Georgian pipeline. Although this measure increased the daily electricity ration from one hour to two hours, long-term fuel increases depended on additional negotiations and of the payment of Armenia's substantial debt to Turkmenistan. In January the State Duma, the lower house of Russia's legislative body, was considering a major grant of credit to Armenia, which would be used in reopening the Armenian Atomic Power Station at Metsamor. The arrangement would be a major step in solidifying economic ties with Russia, which has also given technical assistance for the plant.

According to Armenian Ministry of Industry figures, 40 percent of the country's industrial 1994 output, worth a total of US$147 million, was sold for hard currency. Among the main customers were Iran, Syria, the United Arab Emirates, Cyprus, Belgium, and several North African countries. Although machinebuilding industries did not work at full capacity in 1994 because of a reduced market in Russia, industry was buoyed by the resumption of full production at the Nairit Chemical Plant after several years of shutdown. Nairit was expected to produce goods worth US$60 million per month in 1995.

Armenia's state commission for privatization vouchers began voucher distribution to the public in October 1994. At that point, vouchers for ten enterprises were available, with another fifty due for consideration in February 1995. High profitability was the chief criterion for listing enterprises for privatization. The Nairit plant and the Armenian Electrical Machine Plant, Armenia's largest and most profitable industrial facilities, were converted to private joint-stock enterprises in January 1995.

In Azerbaijan, hopes for economic improvement depended most on foreign investment in offshore oil deposits in the Caspian Sea. Those hopes were subdued somewhat by disagreements over the September 1994 agreement of Western, Russian, and Iranian oil interests to aid Socar, Azerbaijan's state oil company, to develop offshore deposits in the Caspian Sea.

Throughout the last months of 1994, Russia insisted that its 10 percent share of the new deal was unfair on the grounds that all Caspian countries should have equal access to Caspian resources. Russia also continued strong opposition to a new pipeline through Iran to Turkey, which the Western partners favored. The Western firms were dismayed by Azerbaijan's offer of 25 percent of its oil deal to Iran, by the political uncertainty that seemed to escalate in Azerbaijan after the oil deal was signed, and by the rapid deterioration of existing Caspian fields, many of which were deserted in early 1995. Experts agreed an important determinant of Azerbaijan's profit from the agreement would be the maintenance of world oil prices.

In December 1994, Russia's military occupation of its separatist Chechen Autonomous Republic closed the main rail line from Russia, the chief trade route to other CIS republics and elsewhere. Replacement trade routes were sought through Iran, Turkey, and the United Arab Emirates. At the same time, hyperinflation continued (the value of the manat had dropped to 4,300 per US$1 at the end of 1994, down from 120 manats per US$1 in October 1993), spurred by full liberalization of prices to conform with IMF credit requirements. The 1995 budget deficit equaled 20 percent of the gross domestic product (GDP--see Glossary). Foreign credit, especially loans from Turkey, was being used to provide food and social services--needs exacerbated by the continuing influx of Karabakh refugees. Economic reform, meanwhile, was delayed by more immediate concerns. Most industries were operating at about 25 percent of capacity in the winter of 1994-95.

In the last months of 1994, Russia struggled to maintain influence in Azerbaijan. Its position was threatened by approval of the multinational Caspian oil deal in September and by the Azerbaijani perception that the West was restraining Armenian aggression in Karabakh. In November President Aliyev met with Russia's President Yeltsin, who offered 300,000 tons of Russian grain and the reopening of Russian railroad lines in an apparent effort to optimize Russia's influence throughout the Transcaucasus. Azerbaijani opposition parties, led by the Azerbaijani Popular Front (APF), continued to predict that Aliyev's overtures to Russia would return Russia to a dominant position in Azerbaijani political and economic affairs. Experts predicted, however, that Russia would continue to play a vital economic role; at the end of 1994, about 60 percent of Azerbaijan's trade turnover involved Russia.

In early 1995, the issue of Nagorno-Karabakh's status continued to stymie the peace talks jointly sponsored in Moscow by the OSCE and Russia under the Budapest agreement of November 1994. Although Azerbaijan had signed several agreements with Nagorno-Karabakh as a full participant, the extent of the region's autonomy remained a key issue, as did the terms of the liberation of Azerbaijan's Lachin and Shusha regions from Armenian occupation. The Azerbaijani position was that the principals of the negotiations were Armenia and Azerbaijan, with the respective Armenian and Azerbaijani communities in Nagorno- Karabakh as "interested parties." (At the end of 1994, an estimated 126,000 Armenians and 37,000 Azerbaijanis remained in the region.) Azerbaijan lodged an official protest against Russian insistence that the Karabakh Armenians constituted a third principal. In February presidents Aliyev and Ter-Petrosian met with presidents Nursultan Nazarbayev of Kazakhstan and Shevardnadze of Georgia in Moscow and expressed optimism that the nine-month cease-fire would hold until complete settlement could be reached. Nazarbayev and the presidents of Russia and Ukraine offered to be guarantors of stability in Nagorno-Karabakh if Azerbaijan would guarantee the region's borders.

After the unsuccessful coup against him by Prime Minister Suret Huseynov in October 1994, Azerbaijan's President Heydar Aliyev maintained his position. Despite loud opposition from the APF and other parties, Aliyev appeared to occupy a strong position at the beginning of 1995. In early 1995, friction developed between Aliyev and Rusul Guliyev, speaker of the Melli-Majlis (National Council), each accusing the other of responsibility for worsening socioeconomic conditions. Former president Abulfaz Elchibey of the APF remained a vocal critic of Aliyev and had a substantial following.

In Georgia, the unresolved conflict with the Abkhazian Autonomous Republic remained the most important issue. The repatriation of Georgian refugees to Abkhazia, a process conducted very slowly by Abkhazian authorities in the early autumn of 1994, ended completely between November 1994 and January 1995. Opposition parties in Georgia, especially the National Liberation Front led by former Prime Minister Tengiz Sigua, increased their pressure on the government to take action, likening Abkhazia to Russia's secessionist Chechen Autonomous Republic, which Russia invaded in December 1994. (In fact, the official position of the Shevardnadze government supported the Russian move, both because of the parallel with Abkhazia and because of the need for continued Russian military monitoring of the cease-fire.) In January an attempted march of 1,400 armed Georgian refugees into Abkhazia was halted by Georgian government troops, and organizer Tengiz Kitovani, former minister of defense, was arrested for having organized the group. Although the UN adopted resolutions in January condemning the Abkhazian refugee policy, UN officials saw little hope of a rapid change in the situation in 1995.

The issue of human rights continued to dog the Shevardnadze administration in late 1994 and early 1995. In February 1995, the Free Media Association of Georgia, which included most of the country's largest independent newspapers, officially protested police oppression and confiscation of newspapers. Newspaper production had already been restricted since the beginning of winter because of Georgia's acute energy shortage.

The Georgian political world was shocked by the assassination in December 1994 of Gia Chanturia, leader of the moderate opposition National Democratic Party and one of the country's most popular politicians. Responsibility for the act was not established. Chanturia's death escalated calls for resignation of the Cabinet of Ministers, an outcome made more likely by the parliament's failure to pass Shevardnadze's proposed 1995 budget and by continued factionalism within the cabinet.

An important emerging figure was Minister of Defense Vardiko Nadibaidze, an army general entrusted in 1994 with finally developing a professional Georgian military force that would reduce reliance on outside forces (such as Russia's) to protect national security. At the end of 1994, Georgian forces were estimated at 15,000 ground troops, 3,000 air and air defense personnel, and 1,500 to 2,000 in the coastal defense force.

Economic reform continued unevenly under the direction of Vice Premier for Economics Temur Basilia. By design, inflation and prices continued to rise in the last months of 1994, and rubles and dollars remained the chief currency instead of the Georgian coupon. In a November 1994 poll, one-third of respondents said they spent their entire income on food. Distribution of privatization vouchers among the population was scheduled to begin in mid-1995. In November 1994, more than 1,500 enterprises had been privatized, most of them classified as commercial or service establishments. A group of Western and Japanese donors pledged a minimum of US$274 million in credits to Georgia in 1995, with another US$162 million available pending "visible success" in economic reform.

In Geneva, peace talks between the Georgian government and the Abkhazian Autonomous Republic reached the eighteen-month mark; the major points of disagreement continued to be the political status of Abkhazia and the repatriation of Georgian refugees. The Abkhazian delegation insisted on equal status with Georgia in a new confederation. The Russian and UN mediators proposed a federal legislature and joint agencies for foreign policy, foreign trade, taxation, energy, communications, and human rights, providing Abkhazia substantially more autonomy than it had had when Georgia became independent but leaving open the question of relative power within such a system. In early February 1995, preliminary accord was reached on several points of the mediators' proposal.

As 1995 began, prospects for stability in the Transcaucasus were marginally better than they had been since the three countries achieved independence in 1991. Much depended on continued strong leadership from presidents Aliyev, Shevardnadze, and Ter-Petrosian, on a peaceful environment across the borders in Russia and Iran, and on free access to the natural resources needed to restart the national economies.

Religious medallion bearing likeness of Saint George

COUNTRY PROFILE

Country

Formal Name
Republic of Georgia.

Short Name
Georgia.

Term for Citizens
Georgian(s).

Capital
Tbilisi.

Date of Independence
April 9, 1991.

Geography

Size
Approximately 69,875 square kilometers.

Topography
Extremely varied; Greater Caucasus and Lesser Caucasus ranges dominate northern and eastern regions. Many rivers flow through mountain gorges into Black Sea and Caspian Sea. Narrow lowland area along Black Sea. Plains region in east.

Climate
Subtropical, humid along coast. Mountains protect from northern influences and create temperature zones according to elevation. Eastern plains, isolated from sea, have continental climate. Year-round snow in highest mountains.

Society

Population
Mid-1994 estimate 5,681,025. Annual growth rate 0.81 percent in 1994. Density seventy-nine per square kilometer in 1994.

Ethnic Groups
In early 1990s, Georgians 70.1 percent, Armenians 8.1 percent, Russians 6.3 percent, Azerbaijanis 5.7 percent, Ossetians 3 percent, and Abkhaz 1.8 percent.

Languages

In early 1990s, official language, Georgian, spoken by 71 percent of population. Russian spoken by 9 percent, followed by Armenian with 7 percent and Azerbaijani with 6 percent.

Religion

In 1993 Georgian Orthodox 65 percent, Muslim 11 percent, Russian Orthodox 10 percent, and Armenian Apostolic 8 percent.

Education and Literacy

Free and compulsory through secondary school. Previous Soviet system modified to eliminate ideology and strengthen Georgian language and history. Some teaching continues in minority languages. Nineteen institutions of higher learning. Literacy estimated at 100 percent by 1980s.

Health

Universal free health care, among best systems in Soviet period, but under severe stress after 1991. Reform program blocked by civil war and political instability in early 1990s. Facilities overtaxed by refugee and emergency care requirements.

Economy

Gross National Product (GNP)

Estimated at US$4.7 billion in 1992, or approximately US$850 per person. Economic growth negative in early 1990s because of destruction of infrastructure, unavailability of inputs, and failure of economic reorganization.

Agriculture

Very productive with irrigation of western lowlands, but efficiency hindered by post-Soviet misallocation of land and materials. Tea and citrus fruits produced in subtropical areas; also grain, sugar beets, fruits, wine, cattle, pigs, and sheep. Over half of cultivated land privatized as of end of 1993.

Industry and Mining

Industry heavily dependent on inputs from other Commonwealth of Independent States (CIS) republics and from abroad. Main products semifinished metals, vehicles, textiles, and chemicals. Coal, copper, and manganese principal minerals.

Energy

Scant domestic fuel reserves; 95 percent imported (mostly oil and natural gas) in 1990. Coal output dropped sharply through early 1990s. Hydroelectric potential high, but mainly untapped. Power output does not meet domestic needs.

Exports

Estimated at US$32.6 million in 1992. Major exports citrus fruits, tea, machinery, ferrous and nonferrous metals, and textiles. Main markets Armenia, Azerbaijan, Bulgaria, Czechoslovakia, Germany, Poland, Russia, and Turkey.

Imports

Estimated at US$43.8 million in 1992. Major imports machinery and parts, fuels, transportation equipment, and textiles. Main suppliers Bulgaria, Czechoslovakia, Poland, Russia, and Ukraine.

Balance of Payments

Estimated as US$23.7 million deficit in 1992.

Exchange Rate

Coupon introduced in early 1993. November 1994 exchange rate 1,625,000 coupons per US$1.

Inflation

Estimated in January 1993 at 50 percent monthly.

Fiscal Year

Calendar year.

Fiscal Policy

Centralized decision making, but large underground economy limits economic control. Extensive manipulation of tax structure in 1992-93 to shrink large budget deficits. Deficits remained high as revenue estimates fell short. Enterprise privatization slow.

Transportation and Telecommunications

Highways

In 1990 about 35,100 kilometers of roads, of which 31,200 hard-surface. Four main highways radiate from Tbilisi, roughly in the cardinal directions, to Russia, Azerbaijan, Armenia, and Black Sea. Tbilisi hub of Caucasus region's highway system.

Railroads

1,421 kilometers of track in 1993. Main links with Russia, Azerbaijan, and Armenia. Substantial disruption in 1992-93 by civil war and fuel shortages. Tbilisi hub of Caucasus region's rail transport.

Civil Aviation

National airline, Orbis, provides direct flights from Tbilisi to some West European cities. Passenger and cargo service limited by fuel shortages in 1991-94. Nineteen of twenty-six airports with permanent-surface runways in 1993; longest runway, at Novoalekseyevka near Tbilisi, about 2,500 meters.

Inland Waterways

None navigable by commercial shipping.

Ports

Batumi, Poti, and Sukhumi on Black Sea, with international shipping connections to other Black Sea ports and Mediterranean ports.

Pipelines

In 1992 approximately 370 kilometers of pipeline for crude oil, 300 kilometers for refined products, and 440 kilometers for natural gas. Subject to disruption.

Telecommunications

About 672,000 telephone lines in use in 1991, twelve per 100 persons; long waiting list for installation. International links overland to CIS countries and Turkey; lowcapacity satellite earth station in operation. Three television stations and numerous radio stations broadcast in Georgian and Russian.

Government and Politics

Government

Two autonomous republics, Abkhazian Autonomous Republic and Ajarian Autonomous Republic; one autonomous region, South Ossetian Autonomous Region. Strong executive (head of state, who is also chairman of parliament) with extensive emergency powers in civil war period of 1992-93. Cabinet of Ministers selected by head of state; power of prime minister secondary to that of head of state. Unicameral parliament (Supreme Soviet, 225 deputies) elects head of state and has legislative power, but is plagued by disorder and fragmentation. Judicial branch, weak in communist era, under reform in early 1990s.

Politics

Twenty-six parties represented in parliament in 1993, of which most seats held by Peace Bloc, October 11 Bloc, Unity Bloc, Green Party, and National Democratic Party. Shifting coalitions back individual programs. Reform slowed by influence of former communists, gradually dispersing. Union of Citizens of Georgia formed in November 1993 to support Eduard Shevardnadze government programs. Shevardnadze remained most popular politician in late 1994.

Foreign Relations

In 1992-94 wide diplomatic campaign to establish relations with Commonwealth of Independent States (CIS) nations, other neighbors, and the West after isolation created by Zviad Gamsakhurdia government in 1991. Balance maintained between warring Armenia and Azerbaijan. Joined CIS in October 1993 after refusing to do so at first, to ensure Russian aid in ending civil war.

International Agreements and Memberships

Member of United Nations, Conference on Security and Cooperation in Europe, International Monetary Fund, European Bank for Reconstruction and Development, and International Bank for Reconstruction and Development.

National Security

Armed Forces

Defense policy made by Council for National Security and Defense, chaired by head of state. Main forces-- National Guard (15,000 troops) and paramilitary Rescue Corps (about 1,000 troops formerly known as the Mkhedrioni)--not fully under government control in 1994. Plans call for national force of 20,000 with two-year compulsory service. About 15,000 Russian troops remained in mid-1993, supplemented in fall of 1993 to prevent widening of civil war and to guard borders. In 1993 Georgia joined CIS mutual security agreements.

Major Military Units

Emphasis in early 1990s on establishing national ground forces, with small air force using training aircraft. Most equipment obtained from Soviet (later Russian) occupation forces--both legally, under official 1992 quota agreement, and illegally.

Military Budget

In 1992 estimated at US$23.6 million, or 8.3 percent of budgeted expenditures.

Internal Security

Since 1992 intelligence operations under Information and Intelligence Service, chaired by head of state. Ministry of Internal Affairs combined security agencies in 1993. Government police authority uneven; white-collar and highway crime rampant in some regions.

HISTORICAL BACKGROUND

Georgia's location at a major commercial crossroads and among several powerful neighbors has provided both advantages and disadvantages through some twenty-five centuries of history. Georgia is comprised of regions having distinctive traits. The ethnic, religious, and linguistic characteristics of the country as a unit coalesced to a greater degree than before under Russian rule in the nineteenth century. Then, beneath a veneer of centralized economic and political control imposed during seventy years of Soviet rule, Georgian cultural and social institutions survived, thanks in part to Georgia's relative distance from Moscow. As the republic entered the post-Soviet period in the 1990s, however, the prospects of establishing true national autonomy based on a common heritage remained unclear.

Figure 4. Georgia, 1994

Although Saint George is the country's patron saint, the name *Georgia* derives from the Arabic and Persian words, *Kurj* and *Gurj*, for the country. In 1991 Georgia-- called *Sakartvelo* in Georgian and *Gruziia* in Russian--had been part of a Russian or Soviet empire almost continuously since the beginning of the nineteenth century, when most of the regions that constitute modern Georgia accepted Russian annexation in order to gain protection from Persia. Prior to that time, some combination of the territories that comprise modern Georgia had been ruled by the Bagratid Dynasty for about 1,000 years, including periods of foreign domination and fragmentation.

Early History

Archeological evidence indicates a neolithic culture in the area of modern Georgia as early as the fifth millennium B.C. Between that time and the modern era, a number of ethnic groups invaded or migrated into the region, merging with numerous indigenous tribes to form the ethnic base of the modern Georgian people. Throughout history the territory comprising the Georgian state varied considerably in size as foreign forces occupied some regions and as centrally ruled federations controlled others.

Christianity and the Georgian Empire

In the last centuries of the pre-Christian era, Georgia, in the form of the kingdom of Kartli-Iberia, was strongly influenced by Greece to the west and Persia to the east. After the Roman Empire completed its conquest of the Caucasus region in 66 B.C., the kingdom was a Roman client state and ally for some 400 years. In A.D. 330, King Marian III's acceptance of Christianity ultimately tied Georgia to the neighboring Byzantine Empire, which exerted a strong cultural influence for several centuries. Although Arabs captured the capital city of Tbilisi in A.D. 645, Kartli-Iberia retained considerable independence under local Arab rulers. In A.D. 813, the Armenian prince Ashot I became the first of the Bagrationi family to rule Georgia. Ashot's reign began a period of nearly 1,000 years during which the Bagratids, as the house was known, ruled at least part of what is now Georgia.

Western and eastern Georgia were united under Bagrat V (r. 1027-72). In the next century, David IV (called the Builder, r. 1099-1125) initiated the Georgian golden age by driving the Turks from the country and expanding Georgian cultural and political influence southward into Armenia and eastward to the

Caspian Sea. That era of unparalleled power and prestige for the Georgian monarchy concluded with the great literary flowering of Queen Tamar's reign (1184-1212). At the end of that period, Georgia was well known in the Christian West (and relied upon as an ally by the Crusaders). Outside the national boundaries, several provinces were dependent to some degree on Georgian power: the Trabzon Empire on the southern shore of the Black Sea, regions in the Caucasus to the north and east, and southern Azerbaijan (see figure 5).

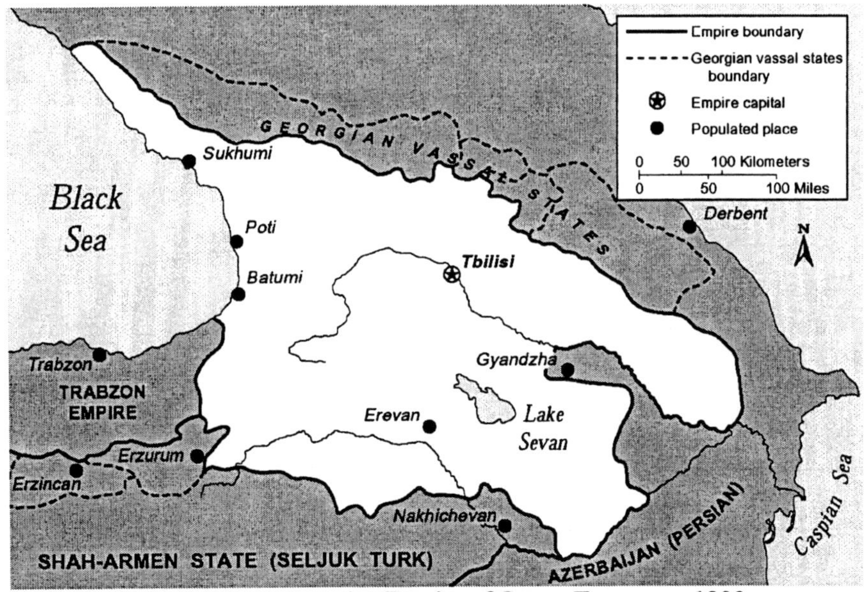

Figure 5. The Georgian Empire of Queen Tamar, ca. 1200
Source: Based on information from Kalistrat Salia, History of the Georgian Nation, Paris, 1983, 182-83.

Occupation and Inclusion in the Russian Empire

The Mongol invasion in 1236 marked the beginning of a century of fragmentation and decline. A brief resurgence of Georgian power in the fourteenth century ended when the Turkic conquerer Timur (Tamerlane) destroyed Tbilisi in 1386. The capture of Constantinople by the Ottoman Turks in 1453 began three centuries of domination by the militant Ottoman and Persian empires, which divided Georgia into spheres of influence in 1553 and subsequently redistributed Georgian territory between them (see figure 6). By the eighteenth century, however, the Bagratid line again had achieved substantial independence under nominal Persian rule. In this period, Georgia was threatened more by rebellious

Georgian and Persian nobles within than by the major powers surrounding the country. In 1762 Herekle II was able to unite the east Georgian regions of Kartli and Kakhetia under his independent but tenuous rule. In this period of renewed unity, trade increased and feudal institutions lost influence in Georgia.

In 1773 Herekle began efforts to gain Russian protection from the Turks, who were threatening to retake his kingdom. In this period, Russian troops intermittently occupied parts of Georgia, making the country a pawn in the explosive Russian-Turkish rivalry of the last three decades of the eighteenth century. After the Persians sacked Tbilisi in 1795, Herekle again sought the protection of Orthodox Russia.

Figure 6. Georgia in the Sixteenth Century
Source: Based on information from Kalistrat Salia, History of the Georgian Nation, Paris, 1983, 253.

Within the Russian Empire

Annexation by the Russian Empire began a new stage of Georgian history, in which security was achieved by linking Georgia more closely than ever with Russia. This subordinate relationship would last nearly two centuries.

Russian Influence in the Nineteenth Century

Because of its weak position, Georgia could not name the terms of protection by the Russian Empire. In 1801 Tsar Alexander I summarily abolished the kingdom of Kartli-Kakhetia, and the heir to the Bagratid throne was forced to abdicate. In the next decade, the Russian Empire gradually annexed Georgia's entire territory. Eastern Georgia (the regions of Kartli and Kakhetia) became part of the Russian Empire in 1801, and western Georgia (Imeretia) was incorporated in 1804. After annexation Russian governors tried to rearrange Georgian feudal society and government according to the Russian model. Russian education and ranks of nobility were introduced, and the Georgian Orthodox Church lost its autocephalous status in 1811. In the second half of the nineteenth century, Russification intensified, as did Georgian rebellions against the process.

Social and Intellectual Developments

By 1850 the social and political position of the Georgian nobility, for centuries the foundation of Georgian society, had deteriorated. A new worker class began to exert social pressure in Georgian population centers. Because the nobility still represented Georgian national interests, its decline meant that the Armenian merchant class, which had been a constructive part of urban life since the Middle Ages, gained greater economic power within Georgia. At the same time, Russian political hegemony over the Caucasus now went unopposed by Georgians. In response to these conditions, Georgian intellectuals borrowed the thinking of Russian and West European political philosophers, forging a variety of theoretical salvations for Georgian nationalism that had little relation to the changing economic conditions of the Georgian people.

By the end of the nineteenth century, Russia, fearing increased Armenian power in Georgia, asserted direct control over Armenian religious and political institutions. In the first decade of the twentieth century, a full-fledged Georgian national liberation movement was led by Marxist followers of the Russian Social Democrat Party. Marxist precepts fell on fertile soil in Georgia; by 1900 migration from rural areas and the growth of manufacturing had generated a fairly cohesive working class led by a new generation of Georgian intellectuals who called for elimination of both the Armenian bourgeoisie and the Russian government bureaucracy. The main foe, however, was tsarist autocracy.

The Spirit of Revolution

In 1905 a large-scale peasant revolt in western Georgia and general strikes in industrial centers throughout the Caucasus caused Russia to declare martial law. As elsewhere in the Russian Empire, the political reforms of 1905 temporarily eased tensions between the Georgian population and the Russian government. For the next decade, the Georgian revolutionaries of the Moscow-based Social Democratic Party were split between the gradualist Menshevik and the radical Bolshevik factions, and the incidence of strikes and mass demonstrations declined sharply between 1906 and 1917. Mensheviks, however, occupied all the Georgian seats in the first two seatings of the Duma, the Russian parliamentary institution established in 1905. In this period, Joseph V. Stalin (a Georgian who changed his name from Ioseb Jugashvili around 1910) became a leader of Bolshevik conspiracies against the Russian government in Georgia and the chief foe of Menshevik leader Noe Zhordania.

World War I and Independence

Because Turkey was a member of the Central Powers in World War I, the Caucasus region became a major battleground in that conflict. In 1915 and 1916, Russian forces pushed southwest into eastern Turkey from bases in the Caucasus, with limited success. As part of the Russian Empire, Georgia officially backed the Allies, although it stood to gain little from victory by either side. By 1916 economic conditions and mass immigration of war refugees had raised social discontent throughout the Caucasus, and the Russian Empire's decade-old experiment with constitutional monarchy was judged a failure.

The revolution of 1917 in Russia intensified the struggle between the Mensheviks and the Bolsheviks in Georgia. In May 1918, Georgia declared its independence under the protection of Germany. Georgia turned toward Germany to prevent opportunistic invasion by the Turks; the move also resulted from Georgians' perception of Germany as the center of European culture. The major European powers recognized Georgia's independence, and in May 1920, Russian leader Vladimir I. Lenin officially followed suit.

To gain peasant support, Zhordania's moderate new Menshevik-dominated government redistributed much of Georgia's remaining aristocratic landholdings to the peasants, eliminating the longtime privileged status of the nobility. The few years of postwar independence were economically disastrous, however, because Georgia did not establish commercial relations with the West, Russia, or its smaller neighbors.

Within the Soviet Union

In seven decades as part of the Soviet Union, Georgia maintained some cultural independence, and Georgian nationalism remained a significant--though at times muted--issue in relations with the Russians. In economic and political terms, however, Georgia was thoroughly integrated into the Soviet system.

The Interwar Years

After independence was declared in 1918, the Georgian Bolsheviks campaigned to undermine the Menshevik leader Zhordania, and in 1921 the Red Army invaded Georgia and forced him to flee. From 1922 until 1936, Georgia was part of a united Transcaucasian Soviet Federated Socialist Republic (TSFSR) within the Soviet Union. In 1936 the federated republic was split up as Armenia, Azerbaijan, and Georgia, which remained separate Soviet socialist republics of the Soviet Union until the end of 1991.

Although Stalin and Lavrenti Beria, his chief of secret police from 1938 to 1953, were both Georgians, Stalin's regime oppressed Georgians as severely as it oppressed citizens of other Soviet republics. The most notable manifestations of this policy were the execution of 5,000 nobles in 1924 as punishment for a Menshevik revolt and the purge of Georgian intellectuals and artists in 1936-37. Another Georgian Bolshevik, Sergo Ordzhonikidze, played an important role in the early 1920s in bringing Georgia and other Soviet republics into a centralized, Moscow-directed state. Ordzhonikidze later became Stalin's top economic official.

World War II and the Late Stalin Period

Georgia was not invaded in World War II. It contributed more than 500,000 fighters to the Red Army, however, and was a vital source of textiles and munitions. Stalin's successful appeal for patriotic unity eclipsed Georgian nationalism during the war and diffused it in the years following. Restoration of autonomy to the Georgian Orthodox Church in 1943 facilitated this process.

The last two decades of Stalin's rule saw rapid, forced urbanization and industrialization, as well as drastic reductions in illiteracy and the preferential treatment of Georgians at the expense of ethnic minorities in the republic. The full Soviet centralized economic planning structure was in place in Georgia by 1934. Between 1940 and 1958, the republic's industrial output grew by 240 percent. In that time, the influence of traditional village life decreased significantly for a large part of the Georgian population.

Post-Stalin Politics

Upon Stalin's death in 1953, Georgian nationalism revived and resumed its struggle against dictates from the central government in Moscow. In the 1950s, reforms under Soviet leader Nikita S. Khrushchev included the shifting of economic authority from Moscow to republic-level officials, but the Russian Khrushchev's repudiation of Stalin set off a backlash in Georgia. In 1956 hundreds of Georgians were killed when they demonstrated against Khrushchev's policy of de-Stalinization. Long afterward many Stalin monuments and place-names--as well as the museum constructed at Stalin's birthplace in the town of Gori, northwest of Tbilisi--were maintained. Only with Mikhail S. Gorbachev's policy of *glasnost* (see Glossary) in the late 1980s did criticism of Stalin become acceptable and a full account of Stalin's crimes against his fellow Georgians become known in Georgia.

Between 1955 and 1972, Georgian communists used decentralization to become entrenched in political posts and to reduce further the influence of other ethnic groups in Georgia. In addition, enterprising Georgians created factories whose entire output was "off the books" (see The Underground Economy). In 1972 the long-standing corruption and economic inefficiency of Georgia's leaders led Moscow to sponsor Eduard Shevardnadze as first secretary of the Georgian Communist Party. Shevardnadze had risen through the ranks of the Communist Youth League (Komsomol) to become a party first secretary at the district level in 1961. From 1964 until 1972, Shevardnadze oversaw the Georgian police from the Ministry of Internal Affairs, where he made a reputation as a competent and incorruptible official.

The First Shevardnadze Period

As party first secretary, Shevardnadze used purges to attack the corruption and chauvinism for which Georgia's elite had become infamous even among the corrupt and chauvinistic republics of the Soviet Union. Meanwhile, a small group of dissident nationalists coalesced around academician Zviad Gamsakhurdia, who stressed the threat that Russification presented to the Georgian national identity. This theme would remain at the center of Georgian-Russian relations into the new era of Georgian independence in the 1990s. Soviet power and Georgian nationalism clashed in 1978 when Moscow ordered revision of the constitutional status of the Georgian language as Georgia's official state language. Bowing to pressure from street demonstrations, Moscow approved Shevardnadze's reinstatement of the constitutional guarantee the same year.

In the 1970s and early 1980s, Shevardnadze successfully walked a narrow line between the demands of Moscow and the Georgians' growing desire for national autonomy. He maintained political and economic control while listening carefully to popular demands and making strategic concessions. Shevardnadze dealt with nationalism and dissent by explaining his policies to hostile audiences and seeking compromise solutions. The most serious ethnic dispute of Shevardnadze's tenure arose in 1978, when leaders of the Abkhazian Autonomous Republic threatened to secede from Georgia, alleging unfair cultural, linguistic, political, and economic restrictions imposed by Tbilisi. Shevardnadze took a series of steps to diffuse the crisis, including an affirmative action program that increased the role of Abkhazian elites in running "their" region, despite the minority status of their group in Abkhazia.

Shevardnadze initiated experiments that foreshadowed the economic and political reforms that Gorbachev later introduced into the central Soviet system. The Abasha economic experiment in agriculture created new incentives for farmers similar to those used in the Hungarian agricultural reform of the time. A reorganization in the seaport of Poti expanded the role of local authorities at the expense of republic and all-union ministries. By 1980 Shevardnadze had raised Georgia's industrial and agricultural production significantly and dismissed about 300 members of the party's corrupt hierarchy. When Shevardnadze left office in 1985, considerable government corruption remained, however, and Georgia's official economy was still weakened by an extensive illegal "second economy." But his reputation for honesty and political courage earned Shevardnadze great popularity among Georgians, the awarding of the Order of Lenin by the Communist Party of the Soviet Union in 1978, and appointment as minister of foreign affairs of the Soviet Union in 1985.

Patiashvili

Jumber Patiashvili, a nondescript party loyalist, succeeded Shevardnadze as head of the Georgian Communist Party. Under Patiashvili, most of

Shevardnadze's initiatives atrophied, and no new policy innovations were undertaken. Patiashvili removed some of Shevardnadze's key appointees, although he could not dismiss his predecessor's many middle-echelon appointees without seriously damaging the party apparatus.

In dealing with dissent, Patiashvili, who distrusted radical and unofficial groups, returned to the usual confrontational strategy of Soviet regional party officials. The party head met major resistance when he backed a plan for a new Transcaucasian railroad that would cut a swath parallel to the Georgian Military Highway in a historic, scenic, and environmentally significant region. In a televised speech, Patiashvili called opponents of the project "enemies of the people"--a phrase used in the 1930s to justify liquidation of Stalin's real and imagined opponents. By isolating opposition groups, Patiashvili forced reformist leaders into underground organizations and confrontational behavior.

In Georgia Gorbachev's simultaneous policies of *glasnost* and continued control energized the forces of nationalism, which pushed the republic out of the central state before the Soviet Union fell apart. The first years of independence were marked by struggle among Georgians for control of the government and by conflict with ethnic minorities seeking to escape the control of Tbilisi.

After Communist Rule

In Georgia Gorbachev's simultaneous policies of *glasnost* and continued control energized the forces of nationalism, which pushed the republic out of the central state before the Soviet Union fell apart. The first years of independence were marked by struggle among Georgians for control of the government and by conflict with ethnic minorities seeking to escape the control of Tbilisi.

Old salt baths and Narikala Fortress, Tbilisi
Courtesy Michael W. Serafin

Old Tbilisi seen from Mtkvari (Kura) River
Courtesy Monica O'Keefe, United States Information Agency

Nationalism Rises

In April 1989, Soviet troops broke up a peaceful demonstration at the government building in Tbilisi. Under unclear circumstances, twenty Georgians, mostly women and children, were killed. The military authorities and the official media blamed the demonstrators, and opposition leaders were arrested. The Georgian public was outraged. What was afterwards referred to as the April Tragedy fundamentally radicalized political life in the republic. Shevardnadze was sent to Georgia to restore calm. He arranged for the replacement of Patiashvili by Givi Gumbaridze, head of the Georgian branch of the Committee for State Security (Komitet gosudarstvennoi bezopasnosti--KGB).

In an atmosphere of renewed nationalist fervor, public opinion surveys indicated that the vast majority of the population was committed to immediate independence from Moscow. Although the communist party was discredited, it continued to control the formal instruments of power. In the months following the April Tragedy, the opposition used strikes and other forms of pressure to undermine communist power and set the stage for de facto separation from the Soviet Union.

The Rise of Gamsakhurdia

Partly as a result of the conspiratorial nature of antigovernment activity prior to 1989, opposition groups tended to be small, tightly knit units organized around prominent individuals. The personal ambitions of opposition leaders prevented the emergence of a united front, but Zviad Gamsakhurdia, the most widely honored and recognized of the nationalist dissidents, moved naturally to a position of leadership. The son of Georgia's foremost contemporary novelist, Gamsakhurdia had gained many enemies during the communist years in acrimonious disputes and irreconcilable factional splits.

Opposition pressure resulted in an open, multiparty election in October 1990. Despite guarantees written into the new law on elections, many prominent opposition parties boycotted the vote, arguing that their groups could not compete fairly and that their participation under existing conditions would only legitimize continuation of Georgia's "colonial status" within the Soviet system.

As an alternative, the opposition parties had held their own election, without government approval, in September 1990. Although the minimum turnout for a valid election was not achieved, the new "legislative" body, called the Georgian National Congress, met and became a center of opposition to the government chosen in the official October election. In the officially sanctioned voting,

Gamsakhurdia's Round Table/Free Georgia coalition won a solid majority in the Supreme Soviet, Georgia's official parliamentary body.

Arguably the most virulently anticommunist politician ever elected in a Soviet republic, Gamsakhurdia was intolerant of all political opposition. He often accused his opposition of treason or involvement with the KGB. The quality of political debate in Georgia was lowered by the exchange of such charges between Gamsakhurdia and opposition leaders such as Gia Chanturia of the National Democratic Party.

After his election, Gamsakhurdia's greatest concern was the armed opposition. Both Gamsakhurdia's Round Table/Free Georgia coalition and some opposition factions in the Georgian National Congress had informal military units, which the previous, communist Supreme Soviet had legalized under pressure from informal groups. The most formidable of these groups were the Mkhedrioni (horsemen), said to number 5,000 men, and the socalled National Guard. The new parliament, dominated by Gamsakhurdia, outlawed such groups and ordered them to surrender their weapons, but the order had no effect. After the elections, independent military groups raided local police stations and Soviet military installations, sometimes adding formidable weaponry to their arsenals. In February 1991, a Soviet army counterattack against Mkhedrioni headquarters had led to the imprisonment of the Mkhedrioni leader.

Gamsakhurdia moved quickly to assert Georgia's independence from Moscow. He took steps to bring the Georgian KGB and Ministry of Internal Affairs (both overseen until then from Moscow) under his control. Gamsakhurdia refused to attend meetings called by Gorbachev to preserve a working union among the rapidly separating Soviet republics. Gamsakhurdia's communications with the Soviet leader usually took the form of angry telegrams and telephone calls. In May 1991, Gamsakhurdia ended the collection in Georgia of Gorbachev's national sales tax on the grounds that it damaged the Georgian economy. Soon Georgia ceased all payments to Moscow, and the central government took steps to isolate the republic economically.

Rather than consent to participate in Gorbachev's March 1991 referendum on preserving a federation of Soviet republics, Gamsakhurdia organized a separate referendum on Georgian independence. The measure was approved by 98.9 percent of Georgian voters. Shortly thereafter, on the second anniversary of the April Tragedy (April 9, 1991), the Georgian parliament passed a declaration of independence from the Soviet Union. Once the Soviet Union collapsed at the end of 1991, Georgia refused to participate in the formation or subsequent activities of the Commonwealth of Independent States (CIS--see Glossary), the loose confederation of independent republics that succeeded the Soviet Union.

The Struggle for Control

In May 1991, Gamsakhurdia was elected president of Georgia (receiving over 86 percent of the vote) in the first popular presidential election in a Soviet republic. Apparently perceiving the election as a mandate to run Georgia personally, Gamsakhurdia made increasingly erratic policy and personnel decisions in the months that followed, while his attitude toward the opposition became more strident. After intense conflict with Gamsakhurdia, Prime Minister Tengiz Sigua resigned in August 1991.

The August 1991 coup attempt against Gorbachev in Moscow marked a turning point in Georgian as well as in Soviet politics. Gamsakhurdia made it clear that he believed the coup, headed by the Soviet minister of defense and the head of the KGB, was both inevitable and likely to succeed. Accordingly, he ordered Russian president Boris N. Yeltsin's proclamations against the coup removed from the streets of Tbilisi. Gamsakhurdia also ordered the National Guard to turn in its weapons, disband, and integrate itself into the forces of the Ministry of Internal Affairs. Opposition leaders immediately denounced this action as capitulation to the coup. In defiance of Gamskhurdia, National Guard commander Tengiz Kitovani led most of his troops out of Tbilisi.

The opposition to Gamsakhurdia, now joined in an uneasy coalition behind Sigua and Kitovani, demanded that Gamsakhurdia resign and call new parliamentary elections. Gamsakhurdia refused to compromise, and his troops forcibly dispersed a large opposition rally in Tbilisi in September 1991. Chanturia, whose National Democratic Party was one of the most active opposition groups at that time, was arrested and imprisoned on charges of seeking help from Moscow to overthrow the government.

In the ensuing period, both the government and extraparliamentary opposition intensified the purchase and "liberation" of large quantities of weapons--mostly from Soviet military units stationed in Georgia--including heavy artillery, tanks, helicopter gunships, and armored personnel carriers. On December 22, intense fighting broke out in central Tbilisi after government troops again used force to disperse demonstrators. At this point, the National Guard and the Mkhedrioni besieged Gamsakhurdia and his supporters in the heavily fortified parliament building. Gunfire and bombs severely damaged central Tbilisi, and Gamsakhurdia fled the city in early January 1992 to seek refuge outside Georgia.

Church and fortress on Georgian Military Highway at Ananuri
Courtesy Gordon Snider

The Military Council

A Military Council made up of Sigua, Kitovani, and Mkhedrioni leader Jaba Ioseliani took control after Gamsakhurdia's departure. Shortly thereafter, a Political Consultative Council and a larger State Council were formed to provide more decisive leadership (see Government and Politics). In March 1992, Eduard Shevardnadze returned to Georgia at the invitation of the Military Council. Shortly thereafter Shevardnadze joined Ioseliani, Sigua, and Kitovani to form the State Council Presidium. All four were given the right of veto over State Council decisions.

Gamsakhurdia, despite his absence, continued to enjoy substantial support within Georgia, especially in rural areas and in his home region of Mingrelia in western Georgia. Gamsakhurdia supporters now constituted another extraparliamentary opposition, viewing themselves as victims of an illegal and unconstitutional putsch and refusing to participate in future elections. Based in the neighboring Chechen Autonomous Republic of Russia, Gamsakhurdia continued to play a direct role in Georgian politics, characterizing Shevardnadze as an agent of Moscow in a neocommunist conspiracy against Georgia. In March 1992, Gamsakhurdia convened a parliament in exile in the Chechen city of Groznyi. In 1992 and 1993, his armed supporters prevented the Georgian government from gaining control of parts of western Georgia.

Threats of Fragmentation

The autonomous areas of South Ossetia and Abkhazia added to the problems of Georgia's post-Soviet governments. By 1993 separatist movements in those regions threatened to tear the republic into several sections. Intimations of Russian interference in the ethnic crises also complicated Georgia's relations with its giant neighbor.

South Ossetia

The first major crisis faced by the Gamsakhurdia regime was in the South Ossetian Autonomous Region, which was largely populated by Ossetians, a separate ethnic group speaking a language based on Persian (see Population and Ethnic Composition). In December 1990, Gamsakhurdia summarily abolished the region's autonomous status within Georgia in response to its longtime efforts to gain independence. When the South Ossetian regional legislature took its first steps toward secession and union with the North Ossetian Autonomous Republic of Russia, Georgian forces invaded. The resulting conflict lasted throughout 1991, causing thousands of casualties and creating tens of thousands of refugees on both sides of the Georgian-Russian border. Yeltsin mediated a cease-fire in July 1992. A year later, the cease-fire was still in place, enforced by Ossetian and Georgian troops together with six Russian battalions. Representatives of the Conference on Security and Cooperation in Europe (CSCE--see Glossary) attempted mediation, but the two sides remained intractable. In July 1993, the South Ossetian government declared negotiations over and threatened to renew large-scale combat, but the cease-fire held through early 1994.

Abkhazia

In the Abkhazian Autonomous Republic of Georgia, the Abkhazian population, like the Ossetians a distinct ethnic group, feared that the Georgians would eliminate their political autonomy and destroy the Abkhaz as a cultural entity. On one hand, a long history of ill will between the Abkhaz and the Georgians was complicated by the minority status of the Abkhaz within the autonomous republic and by periodic Georgianization campaigns, first by the Soviet and later by the Georgian government. On the other hand, the Georgian majority in Abkhazia resented disproportionate distribution of political and administrative positions to the Abkhaz. Beginning in 1978, Moscow had sought to head off Abkhazian demands for independence by allocating as much as 67 percent of party and government positions to the Abkhaz, although, according to the 1989 census, 2.5 times as many Georgians as Abkhaz lived in Abkhazia.

Tensions in Abkhazia led to open warfare on a much larger scale than in South Ossetia. In July 1992, the Abkhazian Supreme Soviet voted to return to the 1925 constitution under which Abkhazia was separate from Georgia. In August 1992, a force of the Georgian National Guard was sent to the Abkhazian capital of Sukhumi with orders to protect Georgian rail and road supply lines, and to secure the border with Russia. When Abkhazian authorities reacted to this transgression of their selfproclaimed sovereignty, hundreds were killed in fighting between Abkhazian and Georgian forces, and large numbers of refugees fled across the border into Russia or into other parts of Georgia. The Abkhazian government was forced to flee Sukhumi.

For two centuries, the Abkhaz had viewed Russia as a protector of their interests against the Georgians; accordingly, the Georgian incursion of 1992 brought an Abkhazian plea for Russia to intervene and settle the issue. An unknown number of Russian military personnel and volunteers also fought on the side of the Abkhaz, and Shevardnadze accused Yeltsin of intentionally weakening Georgia's national security by supporting separatists. After the failure of three cease-fires, in September 1993 Abkhazian forces besieged and captured Sukhumi and drove the remaining Georgian forces out of Abkhazia. In the fall of 1993, mediation efforts by the United Nations (UN) and Russia were slowed by

Georgia's struggle against Gamsakhurdia's forces in Mingrelia, south of Abkhazia. In early 1994, a de facto ceasefire remained in place, with the Inguri River in northwest Georgia serving as the dividing line. Separatist forces made occasional forays into Georgian territory, however.

In September 1993, Gamsakhurdia took advantage of the struggle in Abkhazia to return to Georgia and rally enthusiastic but disorganized Mingrelians against the demoralized Georgian army. Although Gamsakhurdia initially represented his return as a rescue of Georgian forces, he actually included Abkhazian troops in his new advance. Gamsakhurdia's forces took several towns in western Georgia, adding urgency to an appeal by Shevardnadze for Russian military assistance. In mid-October the addition of Russian weapons, supply-line security, and technical assistance turned the tide against Gamsakhurdia and brought a quick end to hostilities on the Mingrelian front (see Foreign Relations). His cause apparently lost, Gamsakhurdia committed suicide in January 1994.

PHYSICAL ENVIRONMENT

Located in the region known as the Caucasus or Caucasia, Georgia is a small country of approximately 69,875 square kilometers--about the size of West Virginia. To the north and northeast, Georgia borders the Russian republics of Chechnya, Ingushetia, and North Ossetia (all of which began to seek autonomy from Russia in 1992). Neighbors to the south are Armenia, Azerbaijan, and Turkey. The shoreline of the Black Sea constitutes Georgia's entire western border (see figure 1).

Topography

Despite its small area, Georgia has one of the most varied topographies of the former Soviet republics (see figure 2). Georgia lies mostly in the Caucasus Mountains, and its northern boundary is partly defined by the Greater Caucasus range. The Lesser Caucasus range, which runs parallel to the Turkish and Armenian borders, and the Surami and Imereti ranges, which connect the Greater Caucasus and the Lesser Caucasus, create natural barriers that are partly responsible for cultural and linguistic differences among regions. Because of their elevation and a poorly developed transportation infrastructure, many mountain villages are virtually isolated from the outside world during the winter.

Earthquakes and landslides in mountainous areas present a significant threat to life and property. Among the most recent natural disasters were massive rock- and mudslides in Ajaria in 1989 that displaced thousands in southwestern Georgia, and two earthquakes in 1991 that destroyed several villages in northcentral Georgia and South Ossetia.

Georgia has about 25,000 rivers, many of which power small hydroelectric stations. Drainage is into the Black Sea to the west and through Azerbaijan to the Caspian Sea to the east. The largest river is the Mtkvari (formerly known by its Azerbaijani name, Kura, which is still used in Azerbaijan), which flows 1,364 kilometers from northeast Turkey across the plains of eastern Georgia, through the capital, Tbilisi, and into the Caspian Sea. The Rioni River, the largest river in western Georgia, rises in the Greater Caucasus and empties into the Black Sea at the port of Poti. Soviet engineers turned the river lowlands along the Black Sea coast into prime subtropical agricultural land, embanked and straightened many stretches of river, and built an extensive system of canals. Deep mountain gorges form topographical belts within the Greater Caucasus.

Climate

Georgia's climate is affected by subtropical influences from the west and mediterranean influences from the east. The Greater Caucasus range moderates local climate by serving as a barrier against cold air from the north. Warm, moist air from the Black Sea moves easily into the coastal lowlands from the west. Climatic zones are determined by distance from the Black Sea and by altitude. Along the Black Sea coast, from Abkhazia to the Turkish border, and in the region known as the Kolkhida Lowlands inland from the coast, the dominant subtropical climate features high humidity and heavy precipitation (1,000 to 2,000 millimeters per year; the Black Sea port of Batumi receives 2,500 millimeters per year). Several varieties of palm trees grow in these regions, where the midwinter average temperature is 5° C and the midsummer average is 22° C.

The plains of eastern Georgia are shielded from the influence of the Black Sea by mountains that provide a more continental climate. Summer temperatures average 20° C to 24° C, winter temperatures 2° C to 4° C. Humidity is lower, and rainfall averages 500 to 800 millimeters per year. Alpine and highland regions in the east and west, as well as a semiarid region on the Iori Plateau to the southeast, have distinct microclimates.

At higher elevations, precipitation is sometimes twice as heavy as in the eastern plains. In the west, the climate is subtropical to about 650 meters; above that altitude (and to the north and east) is a band of moist and moderately warm weather, then a band of cool and wet conditions. Alpine conditions begin at about 2,100 meters, and above 3,600 meters snow and ice are present year-round.

Environmental Issues

Beginning in the 1980s, Black Sea pollution has greatly harmed Georgia's tourist industry. Inadequate sewage treatment is the main cause of that condition. In Batumi, for example, only 18 percent of wastewater is treated before release into the sea. An estimated 70 percent of surface water contains health-endangering bacteria to which Georgia's high rate of intestinal disease is attributed.

The war in Abkhazia did substantial damage to the ecological habitats unique to that region. In other respects, experts considered Georgia's environmental problems less serious than those of more industrialized former Soviet republics. Solving Georgia's environmental problems was not a high priority of the national government in the post-Soviet years, however; in 1993 the minister for protection of the environment resigned to protest this inactivity. In January 1994, the Cabinet of Ministers announced a new, interdepartmental environmental monitoring system to centralize separate programs under the direction of the Ministry of Protection of the Environment. The system would include a central environmental and information and research agency. The Green Party used its small contingent in the parliament to press environmental issues in 1993.

POPULATION AND ETHNIC COMPOSITION

Over many centuries, Georgia gained a reputation for tolerance of minority religions and ethnic groups from elsewhere, but the postcommunist era was a time of sharp conflict among groups long considered part of the national fabric. Modern Georgia is populated by several ethnic groups, but by far the most numerous of them is the Georgians. In the early 1990s, the population was increasing slowly, and armed hostilities were causing large-scale emigration from certain regions. The ethnic background of some groups, such as the Abkhaz, was a matter of sharp dispute.

Population Characteristics

According to the Soviet Union's 1989 census, the total population of Georgia was 5.3 million. The estimated population in 1993 was 5.6 million. Between 1979 and 1989, the population grew by 8.5 percent, with growth rates of 16.7 percent among the urban population and 0.3 percent in rural areas. In 1993 the overall growth rate was 0.8 percent. About 55.8 percent of the population was classified as urban; Tbilisi, the capital and largest city, had more than 1.2 million inhabitants, or approximately 23 percent of the national total. The capital's population grew by 18.1 percent between 1979 and 1989, mainly because of migration from rural areas. Kutaisi, the second largest city, had a population of about 235,000.

In 1991 Georgia's birth rate was seventeen per 1,000 population, its death rate nine per 1,000. Life expectancy was seventy-five years for females and sixty-seven years for males. In 1990 the infant mortality rate was 196 per 10,000 live births. Average family size in 1989 was 4.1, with larger families predominantly located in rural areas. In the 1980s and early 1990s, the Georgian population was aging slowly; the cohort under age nineteen shrank slightly and the cohort over sixty increased slightly as percentages of the entire population during that period. The Georgian and Abkhazian populations were the subjects of substantial international study by anthropologists and gerontologists because of the relatively high number of centenarians among them.

Ethnic Minorities

Regional ethnic distribution is a major cause of the problems Georgia faces along its borders and within its territory (see figure 7). Russians, who make up the third largest ethnic group in the country (6.7 percent of the total population in 1989), do not constitute a majority in any district. The highest concentration of Russians is in Abkhazia, but the overall dispersion of the Russian population restricts political representation of the Russians' interests.

Azerbaijanis are a majority of the population in the districts of Marneuli and Bolnisi, south of Tbilisi on the Azerbaijan border, while Armenians are a majority in the Akhalkalaki, Ninotsminda, and Dmanisi districts immediately to the west of the Azerbaijani-dominated regions and just north of the Armenian border. Despite the proximity and intermingling of Armenian and Azerbaijani populations in Georgia, in the early 1990s few conflicts in Georgia reflected the hostility of the Armenian and Azerbaijani nations over the territory of Nagorno-Karabakh (see

Nagorno-Karabakh and Independence; National Security). Organizations in Georgia representing the interests of the Armenian and Azerbaijani populations had relatively few conflicts with authorities in Tbilisi in the first postcommunist years.

Figure 7. Ethnic Groups in Georgia

Under Soviet rule, a large part of Georgian territory was divided into autonomous regions that included concentrations of non-Georgian peoples. The largest such region was the Abkhazian Autonomous Soviet Socialist Republic (Abkhazian ASSR; after Georgian independence, the Abkhazian Autonomous Republic). The distribution of territory and the past policies of tsarist and Soviet rule meant that in 1989 the Abkhaz made up only 17.8 percent of the population of the autonomous republic named for them (compared with 44 percent Georgians and 16 percent Russians). The Abkhaz constituted less than 2 percent of the total population of Georgia. Although Georgian was the prevailing language of the region as early as the eighth century A.D., Abkhazia was a separate Soviet republic from 1921 until 1930, when it was incorporated into Georgia as an autonomous republic.

In the thirteenth century, Ossetians arrived on the south side of the Caucasus Mountains, in Georgian territory, when the Mongols drove them from what is now the North Ossetian Autonomous Republic of Russia. In 1922 the South Ossetian Autonomous Region was formed within the new Transcaucasian republic of the Soviet Union. The autonomous region was abolished officially by the Georgian government in 1990, then reinstated in 1992. South Ossetia includes many all-Georgian villages, and the Ossetian population is concentrated in the cities of Tskhinvali and Java. Overall, in the 1980s the population in South Ossetia was 66 percent Ossetian and 29 percent Georgian. In 1989 more than 60 percent of the Ossetian population of Georgia lived outside South Ossetia.

The Ajarian Autonomous Soviet Socialist Republic (Ajarian ASSR) in southwest Georgia was redesignated the Ajarian Autonomous Republic in 1992. The existence of that republic reflects the religious and cultural differences that developed when the Ottoman Empire occupied part of Georgia in the sixteenth century and converted the local population to Islam. The Ajarian region was not included in Georgia until the Treaty of Berlin separated it from the Ottoman Empire in 1878. An autonomous republic within Georgia was declared in 1921. Because the Ajarian population is indistinguishable from Georgians in language and belongs to the same ethnic group, it generally considers itself Georgian. Eventually "Ajarian" was dropped from the ethnic categories in the Soviet national census. Thus, in the 1979 census the ethnic breakdown of the region showed about 80 percent Georgians (including Ajars) and 10 percent Russians. Nevertheless, the autonomous republic remains an administrative subdivision of the Republic of Georgia, local elites having fought hard to preserve the special status that this distinction affords them.

The so-called Meskhetian Turks are another potential source of ethnic discord. Forcibly exiled from southern Georgia to Uzbekistan by Stalin during World War II, many of the estimated 200,000 Meskhetian Turks outside Georgia sought to return to their homes in Georgia after 1990. Many Georgians argued that the Meskhetian Turks had lost their links to Georgia and hence had no rights that would justify the large-scale upheaval resettlement would cause. However, Shevardnadze argued that Georgians had a moral obligation to allow this group to return.

Among the leading ethnic groups, the fastest growth between 1979 and 1989 occurred in the Azerbaijani population and the Kurds (see Glossary), whose numbers increased by 20 percent and 30 percent, respectively. This trend worried Georgians, even though both groups combined made up less than 7 percent of the republic's population. Over the same period, the dominant Georgians' share of the population increased from 68.8 percent to 70.1 percent. Ethnic shifts after 1989-- particularly the emigration of Russians, Ukrainians, and Ossetians--were largely responsible for the Georgians' increased share of the population.

LANGUAGE, RELIGION, AND CULTURE

For centuries, Georgia's geographic position has opened it to religious and cultural influences from the West, Persia, Turkey, and Russia. The resultant diversity continues to characterize the cultural and religious life of modern Georgia. However, the Georgian language displays unique qualities that cannot be attributed to any outside influence.

Language

Even more than religion, the issue of language was deeply entwined with political struggles in Georgia under communist rule. As elsewhere, language became a key factor in ethnic selfidentification under the uniformity of the communist system. Written in a unique alphabet that began to exhibit distinctions from the Greek alphabet in the fifth century A.D., Georgian is linguistically distant from Turkic and Indo-European languages. In the Soviet period, Georgians fought relentlessly to prevent what they perceived as the encroachment of Russian on their native language. Even the republic's Soviet-era constitutions specified Georgian as the state language. In 1978 Moscow failed to impose a constitutional change giving Russian equal status with Georgian as an official language when Shevardnadze yielded to mass demonstrations against the amendment (see Within the Soviet Union). Nevertheless, the Russian language predominated in official documents and communications from the central government. In 1991 the Gamsakhurdia government reestablished the primacy of Georgian, to the dismay of minorities that did not use the language. In 1993 some 71 percent of the population used Georgian as their first language. Russian was the first languages of 9 percent, Armenian of 7 percent, and Azerbaijani of 6 percent.

His Holiness Ilia II, Patriarch of Mtskheta and All Georgia, leader of Georgian
Orthodox Church
Courtesy Janet A. Koczak

Religion

The wide variety of peoples inhabiting Georgia has meant a correspondingly
rich array of active religions. The dominant religion is Christianity, and the
Georgian Orthodox Church is by far the largest church. The conversion of the
Georgians in A.D. 330 placed them among the first peoples to accept Christianity.
According to tradition, a holy slave woman, who became known as Saint Nino,
cured Queen Nana of Iberia of an unknown illness, and King Marian III accepted
Christianity when a second miracle occurred during a royal hunting trip. The
Georgians' new faith, which replaced Greek pagan and Zoroastrian beliefs, was to
place them permanently on the front line of conflict between the Islamic and
Christian worlds. As was true elsewhere, the Christian church in Georgia was
crucial to the development of a written language, and most of the earliest written
works were religious texts. After Georgia was annexed by the Russian Empire, the
Russian Orthodox Church took over the Georgian church in 1811. The colorful
frescoes and wall paintings typical of Georgian cathedrals were whitewashed by
the Russian occupiers.

The Georgian church regained its autonomy only when Russian rule ended in 1918. Neither the Georgian Menshevik government nor the Bolshevik regime that followed considered revitalization of the Georgian church an important goal, however. Soviet rule brought severe purges of the Georgian church hierarchy and constant repression of Orthodox worship. As elsewhere in the Soviet Union, many churches were destroyed or converted into secular buildings. This history of repression encouraged the incorporation of religious identity into the strong nationalist movement in twentieth-century Georgia and the quest of Georgians for religious expression outside the official, government-controlled church. In the late 1960s and early 1970s, opposition leaders, especially Zviad Gamsakhurdia, criticized corruption in the church hierarchy. When Ilia II became the patriarch (catholicos) of the Georgian Orthodox Church in the late 1970s, he brought order and a new morality to church affairs, and Georgian Orthodoxy experienced a revival. In 1988 Moscow permitted the patriarch to begin consecrating and reopening closed churches, and a large-scale restoration process began. In 1993 some 65 percent of Georgians were Georgian Orthodox, 11 percent were Muslim, 10 percent Russian Orthodox, and 8 percent Armenian Apostolic.

Non-Orthodox religions traditionally have received tolerant treatment in Georgia. Jewish communities exist throughout the country, with major concentrations in the two largest cities, Tbilisi and Kutaisi. Azerbaijani groups have practiced Islam in Georgia for centuries, as have the Abkhazian and Ajarian groups concentrated in their respective autonomous republics. The Armenian Apostolic Church, whose doctrine differs in some ways from that of Georgian Orthodoxy, has autocephalous status.

The Arts

In many art forms, Georgia has a tradition spanning millennia. The golden age of the Georgian Empire (early twelfth century to early thirteenth century) was the time of greatest development in many forms, and subsequent centuries of occupation and political domination brought decline or dilution. Folk music and dance, however, remain an important part of Georgia's unique culture, and Georgians have made significant contributions to theater and film in the late twentieth century.

Literature

Among literary works written in Georgian, Shota Rustaveli's long poem *The Knight in the Panther Skin* occupies a unique position as the Georgian national epic. Supposedly Rustaveli was a government official during the reign of Queen Tamar (1184- 1212), late in the golden age. In describing the questing adventures of three hero-knights, the poem includes rich philosophical musings that have become proverbs in Georgian. Even during communist rule, the main street of the Georgian capital was named after Rustaveli.

Architecture

Starting in its earliest days, Georgia developed a unique architectural style that is most visible in religious structures dating as far back as the sixth century A.D. The cupola structure typical of Georgian churches probably was based on circular domestic dwellings that existed as early as 3000 B.C. Roman, Greek, and Syrian architecture also influenced this style. Persian occupation added a new element, and in the nineteenth century Russian domination created a hybrid architectural style visible in many buildings in Tbilisi. The so-called Stalinist architecture of the mid-twentieth century also left its mark on the capital.

Painting, Sculpture, and Metalworking

Like literature, Georgian mural painting reached a zenith during the golden age of the twelfth and thirteenth centuries. Featuring both religious and secular themes, many monuments of this and the later Byzantine- and Persian-influenced periods were destroyed by the Russians in the nineteenth century. Examples of Georgian religious painting remain in some of the old churches. Stone carving and metalworking traditions had developed in antiquity, when Roman and Greek techniques were incorporated. In the golden age, sculpture was applied most often to the outside of buildings. In the twentieth century, several Georgian sculptors have gained international recognition. Among them is Elguja Amasukheli, whose monuments are landmarks in Tbilisi. Metalworking was well established in the Caucasus among the ancestors of the Georgians as early as the Bronze Age (second millennium B.C.). This art form, applied to both religious and secular subjects, declined in the Middle Ages.

Music and Dance

Georgia is known for its rich and unique folk dance and music. The Georgian State Dance Company, founded in the 1940s, has traveled around the world performing spectacular renditions of traditional Georgian dances. Unique in folk-dancing tradition, Georgian male performers dance on their toes without the help of special blocked shoes. Georgian folk music, featuring complex, three-part, polyphonic harmonies, has long been a subject of special interest among musicologists. Men and women sing in separate ensembles with entirely different repertoires. Most Georgian folk songs are peculiar to individual regions of Georgia. The inspiration is most often the church, work in the fields, or special occasions. The Rustavi Choir, formed in 1968, is the best known Georgian group performing a traditional repertoire.

In modern Georgia, folk songs are most frequently sung around the table. The ceremonial dinner (*supra*), a frequent occurrence in Georgian homes, is a highly ritualized event that itself forms a direct link to Georgia's past. On such occasions, rounds of standardized and improvised toasts typically extend long into the night. Georgian cuisine, which includes a variety of delicate sauces and sharp spices, is also an important part of the culture that links the generations. In the Soviet period, the best restaurants in the large cities of other republics were often Georgian.

Film and Theaters

In the postwar era, Georgian filmmaking and theater developed an outstanding reputation in the Soviet Union. Several Georgian filmmakers achieved international recognition in this period. Perhaps the single most important film of the *perestroika* (see Glossary) period was Tengiz Abuladze's *Repentance.* This powerful work, which won international acclaim when released in 1987, showed the consequences of Stalin's Great Terror of the 1930s through a depiction of the reign of a fictional local dictator. In 1993, despite chaotic political conditions, Tbilisi hosted the Golden Eagle Film Festival of the Black Sea Basin Countries, Georgia's first international film festival. Georgians also excel in theater. The Tbilisi-based Rustaveli Theater has been acclaimed internationally for its stagings (in Georgian) of the works of William Shakespeare and German dramatist Bertolt Brecht.

EDUCATION, HEALTH, AND WELFARE

In 1992 Georgia retained the basic structure of education, health, and social welfare programs established in the Soviet era, although major reforms were being discussed. Georgia's requests for aid from the West have included technical assistance in streamlining its social welfare system, which heavily burdens the economy and generally fails to help those in greatest need.

Elementary school children in English class, Children's Palace, Tbilisi
Courtesy Janet A. Koczak

Education

In the Soviet era, the Georgian population achieved one of the highest education levels in the Soviet Union. In 1989 some 15.1 percent of adults in Georgia had graduated from a university or completed some other form of higher education. About 57.4 percent had completed secondary school or obtained a specialized secondary education. Georgia also had an extensive network of 230 scientific and research institutes employing more than 70,000 people in 1990. The Soviet system of free and compulsory schooling had eradicated illiteracy by the 1980s, and Georgia had the Soviet Union's highest ratio of residents with a higher or specialized secondary education.

During Soviet rule, the Communist Party of the Soviet Union (CPSU--see Glossary) controlled the operation of the Georgian education system. Theoretically, education was inseparable from politics, and the schools were deemed an important tool in remaking society along Marxist-Leninist lines. Central ministries for primary and secondary education and for higher and specialized education transmitted policy decisions to the ministries in the republics for implementation in local and regional systems. Even at the local level, most administrators were party members. The combination of party organs and government agencies overseeing education at all levels formed a huge bureaucracy that made significant reform impossible. By the mid-1980s, an education crisis was openly recognized everywhere in the Soviet Union.

In the early 1990s, Soviet education institutions were still in place in Georgia, although Soviet-style political propaganda and authoritarian teaching methods gradually disappeared. Most Georgian children attended general school (grades one to eleven), beginning at age seven. In 1988 some 86,400 students were enrolled in Georgia's nineteen institutions of higher learning. Universities are located in Batumi, Kutaisi, Sukhumi, and Tbilisi. In the early 1990s, private education institutes began to appear. Higher education was provided almost exclusively in Georgian, although 25 percent of general classes were taught in a minority language. Abkhazian and Ossetian children were taught in their native language until fifth grade, when they began instruction in Georgian or Russian.

Health

The Soviet system of health care, which embraced all the republics, included extensive networks of state-run hospitals, clinics, and emergency first aid stations. The huge government health bureaucracy in Moscow set basic policies for the entire country, then transmitted them to the health ministries of the republics. In the republics, programs were set up by regional and local health authorities. The emphasis was on meeting national standards and quotas for patient visits, treatments provided, and hospital beds occupied, with little consideration of regional differences or requirements.

Under this system, the average Georgian would go first to one of the polyclinics serving all the residents of a particular area. In the mid-1980s, polyclinics provided about 90 percent of medical care, offering very basic diagnostic services. In addition, most workplaces had their own clinics, which minimized time lost from work for medical reasons. The hospital system provided more complex diagnosis and treatment, although overcrowding often resulted from the admission of patients with minor complaints. Crowding was exacerbated by official standards requiring hospital treatment of a certain duration for every type of complaint.

The Soviet system placed special emphasis on treatment of women and children; many specialized treatment, diagnostic, and advanced-study centers offered pediatric, obstetric, and gynecological care. Maternity services and prenatal care were readily accessible. Emergency first aid was provided by specialized ambulance teams, most of which had only very basic equipment. Severe cases went to special emergency hospitals because regular hospitals lacked emergency rooms. Although this system worked efficiently in urban centers such as Tbilisi, it did not reach remote areas. Most Georgians cared for elderly family members at home, and nursing care was generally mediocre. Georgian health spas were a vital part of the Soviet Union's well-known sanatorium system, access to which was a privilege of employment in most state enterprises.

When the Soviet Union dissolved, it left a legacy of health problems to the respective republics, which faced the necessity of organizating separate health systems under conditions of scarce resources. By 1990 the Soviet health system had become drastically underfunded, and the incidence of disease and accidents was increased by poor living standards and environmental hazards. Nominally equal availability of medical treatment and materials was undermined by the privileged status of elite groups that had access to the country's best medical facilities. In 1990 the former republics also differed substantially in health conditions and availability of care (see table 1). Subsequent membership in the Commonwealth of Independent States, to which Georgia committed itself in late 1993, did not affect this inequality.

Table 1. Armenia, Azerbaijan, and Georgia:
Selected Health and Health Care Statistics, 1989, 1990, and 1991

	Armenia	Azerbaijan	Georgia
Disease diagnosis[1]			
Tuberculosis	17.6	36.2	28.9
Viral hepatitis	279.0	310.5	226.3
Cancer	223.1	224.9	140.9
Hospital beds[2]	89.4	99.4	110.7
Doctors[2]	42.8	38.9	59.2
Pharmacists[3]	7.0	6.7	14.3
Infant mortality[4]	17.1	25.0	15.9

[1] For tuberculosis and cancer, first diagnoses per 100,000 population in 1990; for viral hepatitis, registered cases per 100,000 population in 1989.

[2] Per 10,000 population: in 1990 for Georgia, in 1991 for Armenia and Azerbaijan.

[3] Per 10,000 population in 1989.

[4] Per 1,000 live births: in 1990 for Georgia, in 1991 for Armenia and Azerbaijan.

Source: Based on information from Christopher M. Davis, "Health Care Crisis: The Former Soviet Union," *RFE/RL Research Report* [Munich], 2, No. 40, October 8, 1993, 36.

According to most standard indicators, in 1991 the health and medical care of the Georgian population were among the best in the Soviet Union. The rate at which tuberculosis was diagnosed, 28.9 cases per 100,000 population in 1990, was third lowest, and Georgia's 140.9 cancer diagnoses per 100,000 population in 1990 was the lowest rate among the Soviet republics. Georgia also led in physicians per capita, with 59.2 per 10,000 population, and in dentists per capita. However, hospital bed availability, 110.7 per 10,000 population in 1990, placed Georgia in the bottom half among Soviet republics, and infant mortality, 15.9 per 1,000 live births in 1990, was at the average for republics outside Central Asia.

Although illegal drugs were available and Georgia increasingly found itself on the international drug-trading route in the early 1990s, the drug culture was confined to a small percentage of the population. The relatively high rate of delinquency among Georgian youth, however, was frequently associated with alcohol abuse.

In 1993 the Republic AIDS and Immunodeficiency Center in Tbilisi reported that sixteen cases of acquired immune deficiency syndrome (AIDS) had been detected; five victims were nonGeorgians and were deported. Of the remaining eleven, two had contracted AIDS through drug use and one through a medical procedure. Despite the small number of cases, the AIDS epidemic has caused considerable alarm in the Georgian medical community, which formed a physicians' anti-AIDS association in 1993. The AIDS center, located in a makeshift facility in Tbilisi, conducts AIDS research and oversees testing in twenty-nine laboratories throughout Georgia, stressing efforts among high-risk groups.

Like other former Soviet republics, Georgia began devising health care reform strategies in 1992. Budget expenditures for health increased drastically once the Soviet welfare system collapsed. Theoretical elements of Georgian health reform were compulsory medical insurance, privatization and foreign investment in institutions providing health care, and stronger emphasis on preventive medicine. Little progress was made in the first two years of the reform process, however. In Georgia political instability and civil war have destroyed medical facilities while increasing the need for emergency care and creating a large-scale refugee problem (see Threats of Fragmentation).

Social Security

In 1985 some 47 percent of Georgia's budget went to support the food, health, and education needs of the population. Social services included partial payment for maternity leave for up to eighteen months and unpaid maternity leave for up to three years. State pensions were automatic after twenty years of work for women and twenty-five years for men. As inflation rose in the postcommunist era, however, a large percentage of older Georgians continued working because their pensions could not support them. In 1991 the social security fund--supported mainly by a payroll tax--provided pensions for 1.3 million persons. The fund also paid benefits for sick leave and rest homes, as well as allowances for families with young children.

In 1992 subsidies were in place for basic commodities, pensions, unemployment benefits, and allowances for single mothers and children. At that time, a payroll tax of 3 percent was designated to support the national unemployment fund. Deficits in the social security fund were nominally covered by the state budget, but budget shortfalls elsewhere shifted that responsibility to the banking system. In 1992 increased benefit payments and the decision not to increase the payroll tax eroded the financial base of the fund.

THE ECONOMY

In the Soviet period, Georgia played an important role in supplying food products and minerals and as a center of tourism for the centralized state economy. However, the republic was also heavily dependent on imports to provide products vital to industrial support. In the post-Soviet years, the Georgian economy suffered a major decline because sources of those products were no longer reliable and because political instability limited the economic reorganization and foreign investment that might support an internationalized, free-market economy. The net material product (NMP--see Glossary) already had declined by 5 percent in 1989 and by 12 percent in 1990, after growing at an annual rate of 6 percent between 1971 and 1985. In late 1993, Shevardnadze reported that industrial production had declined by 60.5 percent in 1993 and that the annual inflation rate had reached 2,000 percent, largely as a result of the economic disruption caused by military conflict within Georgia's borders.

Conditions in the Soviet System

Georgian nationalists contended that Georgia's role in the "division of labor" among Soviet republics was unfairly assigned and that other republics, especially Russia, benefited from the terms of trade set by Moscow. Georgian manganese, for example, went to Soviet steel plants at an extremely low price, and Georgian agricultural goods also sold at very low prices in other republics. At the same time, Georgia paid high prices for machinery and equipment purchased elsewhere in the Soviet Union and in Eastern Europe. Despite Georgia's popularity as a tourist destination, the republic reaped few benefits because most hard currency earnings from tourism went to Moscow and because Soviet tourists paid little for their state-sponsored "vacation packages." Georgia, however, benefited from energy prices that were far below world market levels.

Despite the ambiguities of official statistics, all evidence indicates that after 1989 Georgia experienced a disastrous drop in industrial output, real income, consumption, capital investment, and virtually every other economic indicator. For example, official statistics showed a decline in national income of 34 percent in 1992 from 1985 levels.

Obstacles to Development

Several noneconomic factors influenced the broad decline of the Georgian economy that began before independence was declared in 1991. National liberation leaders used strikes in 1989 and 1990 to gain political concessions from the communist leadership, and a 1990 railroad strike, for instance, paralyzed most of the Georgian economy. In 1991 the Gamsakhurdia government ordered strikes at enterprises subordinated to ministries in Moscow as a protest against Soviet interference in South Ossetia (see Ethnic Minorities).

Although combat in Georgia in the period after 1991 left most of the republic unscathed, the economy suffered greatly from military action. Railroad transport between Georgia and Russia was disrupted severely in 1992 and 1993 because most lines from Russia passed through regions of severe political unrest. Georgia's natural gas pipeline to the north entered Russia through South Ossetia and thus was subject to attack during the ethnic war that began in that region in late 1990. In western Georgia, Gamsakhurdia's forces and Abkhazian separatists often stopped trains or blew up bridges in 1992. As a result, supplies could only enter Georgia through the Black Sea ports of Poti and Batumi or over a circuitous route from Russia through Azerbaijan.

In both the Soviet and the post-Soviet periods, conflicts between Georgia and Moscow broke many vital links in the republic's economy. Official 1988 data showed imports to Georgia from other republics of more than 5.2 billion rubles and exports of over 5.5 billion rubles. As a result of Gamsakhurdia's policies, goods destined for Russia were withheld by Georgian officials. The Soviet leadership, encouraged by conservative provincial leaders in the Russian regions bordering Georgia, responded with their own partial economic blockade of Georgia in late 1990 and 1991. All-union enterprises in Georgia stopped receiving most of their supplies from outside the republic. The strangling of energy resources forced much of Georgian industry to shut down in 1991.

The Underground Economy

Economic statistics for Georgia are difficult to evaluate for both the Soviet era and the post-Soviet period, primarily because of the country's large underground economy. Traditional Georgian familial and clan relations have intensified the economic corruption that infused the entire communist system. Local elites in the communist party joined with underground speculators and entrepreneurs to form an economic mafia. Repeated efforts to eradicate this phenomenon, including an aggressive effort by Shevardnadze in the early 1970s, apparently had little impact. In the postcommunist period, struggles for economic control among competing mafias have been an important part of the political conflict plaguing Georgia.

Jewelry-making and gun-repair stand in Tbilisi
Courtesy A. James Firth, United States Department of Agriculture

Wages and Prices

Until 1991 Georgia's price system and inflation rate generally coincided with those of the other Soviet republics. Under central planning, prices of state enterprise products were fixed by direct regulation, fixed markup rates, or negotiation at the wholesale level with subsequent sanction by state authority. The prices of agricultural products from the private sector fluctuated freely in the Soviet system.

Once it forsook the artificial conditions of the Soviet system, Georgia faced the necessity for major changes in its pricing policy. Following the political upheaval of late 1991, which delayed price adjustments, the Georgian government raised the prices of basic commodities substantially in early 1992, to match adjustments made in most of the other former Soviet republics. The price of bread, for example, rose from 0.4 ruble to 4.8 rubles per kilogram. By the end of 1992, all prices except those for bread, fuel, and transportation had been liberalized in order to avoid distortions and shortages. This policy brought steep inflation rates throughout 1993.

Beginning in 1991, a severe shortage of ruble notes restricted enterprises from acquiring enough currency to prevent a significant drop in real wages. In early 1992, public-sector wages were doubled, and every Georgian received an additional 40 rubles per month to compensate for the rising cost of living. Such compensatory increases were far below those in other former Soviet republics, however. In 1992 the Shevardnadze government considered wage indexing or regular adjustment of benefits to the lowest wage groups as ways of improving the public's buying power.

In mid-1993 the majority of Georgians still depended on state enterprises for their salaries, but in most cases some form of private income was necessary to live above the poverty level. Private jobs paid substantially more than state jobs, and the discrepancy grew larger in 1993. For example, in 1993 a secretary in a private company earned the equivalent of US$30 per month while a state university professor made the equivalent of US$4 per month.

Banking, the Budget, and the Currency

In the spring of 1991, Georgian banks ended their relationship with parent banks in Moscow. The National Bank of Georgia was created in mid-1991 as an independent central national bank; its main function was to ensure the stability of the national currency, and it was not responsible for obligations incurred by the government. The National Bank also assumed all debts of Georgian banks to the state banks in Moscow.

In 1992 the national system also included five specialized government commercial banks and sixty private commercial banks. The five government-owned commercial banks provided 95 percent of bank credit going to the economy. They included the Agricultural and Industrial Bank of Georgia, the Housing Bank of Georgia, and the Bank for Industry and Construction, which were the main sources of financing for state enterprises during this period. Private commercial banks, which began operation in 1989, grew rapidly in 1991-92 because of favorable interest rates; new banking laws were passed in 1991 to cover their activity.

Under communist rule, transfers from the Soviet national budget had enabled Georgia to show a budget surplus in most years. When the Soviet contribution of 751 million rubles--over 5 percent of Georgia's gross domestic product (GDP--see Glossary) became unavailable in 1991, the Georgian government ran a budget deficit estimated at around 2 billion rubles. The destruction of government records during the Tbilisi hostilities of late 1991 left the new government lacking reliable information on which to base financial policy for 1992 and beyond (see After Communist Rule).

In 1992 the government assumed an additional 2 billion to 3 billion rubles of unpaid debts from state enterprises, raising the deficit to between 17 and 21 percent of GDP (see table 2). By May 1992, when the State Council approved a new tax system, the budget deficit was estimated at 6 billion to 7 billion rubles. The deficit was exacerbated by military expenditures associated with the conflicts in South Ossetia and Abkhazia and by the cost of dealing with natural disasters.

The 1992 budget was restricted by a delay in the broadening of the country's tax base, the cost of assuming defense and security expenses formerly paid by the Soviet Union, the doubling of state wages, and the cost of earthquake relief in the north. When the 1993 budget was proposed, only 11 billion of the prescribed 43.6 billion rubles of expenditures were covered by revenues.

Tax reform in early 1992 added an excise tax on selected luxury items and a flat-rate value-added tax (VAT--see Glossary) on most goods and services, while abolishing the turnover and sales taxes of the communist system. In 1992 tax revenues fell below the expected level, however, because of noncompliance with new tax requirements; a government study showed that 80 percent of businesses underpaid their taxes in 1992.

Table 2. Georgia: Government Budget, 1991 and 1992
(in millions of rubles)

	1991		1992	
	Budget	Actual	Budget	Approved
Revenues				
Tax revenues	3,972	4,631	22,202	15,218
Nontax revenues	1,295	1,567	2,898	3,655
Total revenues	5,267	6,198	25,100	18,873
Expenditures				
National economy[1]	2,979	2,878	11,885	20,815
Social and cultural				
Education and culture	n.a.	1,520	n.a.	8,421
Health and sports	n.a.	812	n.a.	2,996
Social security	n.a.	120	n.a.	34
Science	n.a.	74	n.a.	874
Total social and cultural	2,717	2,526	10,082	12,325
Administration and law enforcement				
State administration	n.a.	124	n.a.	1,28 8
Internal security and defense	n.a.	118	n.a.	5,282
Total administration and law enforcement	245	242	2,631	6,570
Other	327	276	1,765	2,972
Total expenditures	6,268	5,922	26,363	42,682
Extrabudgetary factors[2]	n.a.	1,000	12,713	-23,320
Interest on foreign debt	0	0	0	21,714
Balance	-1,001	- 724	-13,976	-68,843

n.a.--not available.

[1] Investment by budgetary institutions and transfers to the enterprise sector.

[2] Errors and omissions and extrabudgetary expenditures for social security fund and net lending.

Source: Based on information from World Bank, *Georgia: A Blueprint for Reforms,* Washington, 1993, 120; and *The Europa World Year Book, 1993,* 1, London, 1993, 50.

In early 1993, Georgia remained in the "ruble zone," still using the Russian ruble as the official national currency. Efforts begun in 1991 to establish a separate currency convertible on world markets were frustrated by political and economic instability. Beginning in August 1993, the Central Bank of Russia began withdrawing ruble banknotes; a new unit, designated the coupon (for value of the coupon--see Glossary), became the official national currency after several months of provisional status. Rubles and United States dollars continued to circulate widely, however, especially in large transactions. After the National Bank of Georgia had been establishing weekly exchange rates for two months, the coupon's exchange rate against the United States dollar inflated from 5,569 to 12,629. In September all salaries were doubled, setting off a new round of inflation. By October the rate had reached 42,000 coupons to the dollar.

Industry

In 1990 about 20 percent of Georgia's 1,029 industrial enterprises, including the largest, were directly administered by the central ministries of the Soviet Union. Until 1991 Georgian industry was integrated with the rest of the Soviet economy. About 90 percent of the raw materials used by Georgian light industry came from outside the republic. The Transcaucasian Metallurgical Plant at Rustavi and the Kutaisi Automotive Works, as well as other centers of heavy industry, depended heavily on commercial agreements with the other Soviet republics. The Rustavi plant, for example, could not operate without importing iron ore, most of which it received (and continues to receive) from Azerbaijan. The Kutaisi works depended on other republics for raw materials, machinery, and spare parts. Georgia contributed significantly to Soviet mineral output, particularly of manganese (a component of steel alloy found in the Chiatura and Kutaisi regions in west-central Georgia) and copper.

In the late 1980s, Georgia's main industrial products were machine tools, prefabricated building structures, cast iron, steel pipe, synthetic ammonia, and silk thread. Georgian refineries also processed gasoline ansd diesel fuel from imported crude oil. Georgian industry made its largest contributions to the Soviet Union's total industrial production in wool fabric, chemical fibers, rolled ferrous metals, and metal-cutting machine tools (see table 3).

Table 3. Georgia: Output of Major Industrial Products, 1990, 1991, and 1992 (in thousands of tons unless otherwise specified)

Product	1990	1991	1992*
Beer (in thousands of decaliters)	9,477	6,011	3,288
Cigarettes (in millions)	11,200	9,800	5,100
Cotton fabric (in millions of square meters)	34	17	13
Diesel fuel	658	495	111
Footwear (in millions of pairs)	13	12	3
Heavy oil (mazut)	898	737	189
Machine tools (in units)	1,565	1,417	1,149
Margarine	34	16	2
Motor fuel	399	324	72
Steel	1,316	962	535
Synthetic fibers	32	20	5
Synthetic resins and plastics	40	26	8
Vegetable oil	14	7	0.1
Wine (in thousands of decaliters)	16,283	12,166	7,130

* Estimated.

Source: Based on information from *The Europa World Year Book, 1993,* 1, London, 1993, 1236; and United States, Central Intelligence Agency, *Handbook of International Economic Statistics, 1993,* Washington, 1993, 76.

Energy Resources

The lack of significant domestic fuel reserves made the Georgian economy extremely dependent on neighboring republics, especially Russia, to meet its energy needs. Under the fuel supply conditions of 1994, only further exploitation of hydroelectric power could enhance energy self-sufficiency. In 1990 over 95 percent of Georgia's fuel was imported. For that reason, the collapse of the Soviet Union in 1991 caused an energy crisis and stimulated a search for alternative suppliers.

The harsh winter of 1991-92 increased fuel demand at a time when supply was especially limited. Oil imports were reduced by the conflict between Armenia and Azerbaijan, cold weather curtailed domestic hydroelectric production, and the price of fuel and energy imports from other former Soviet republics rose drastically because of Georgia's independent political stance and the new economic realities throughout the former union. Beginning in December 1991, industries received only about one-third of the energy needed for full-scale operation, and most operated far below capacity throughout 1992.

Hydroelectric station on Georgian Military Highway between Tbilisi and Mtskheta

Small amounts of oil were discovered in the Samgori region (southern Georgia) in the 1930s and in eastern Georgia in the 1970s, but no oil exploration has occurred in most of the republic. In 1993 some 96 percent of Georgia's oil came from Azerbaijan and Russia, although new supply agreements had been reached with Iran and Turkey. Oil and gas pipelines connect Georgia with Azerbaijan, Armenia, Russia, and Turkmenistan. Refinery and storage facilities in Batumi receive oil through a long pipeline from Baku in Azerbaijan.

Coal is mined in Abkhazia and near Kutaisi, but between 1976 and 1991 output fell nearly 50 percent, to about 1 million tons. The largest deposits, both in Abkhazia, are estimated to contain 250 million tons and 80 million tons, respectively. Domestic coal provides half the Rustavi plant's needs and fuels some electrical power generation. In 1993 natural gas, nearly all of which was imported, accounted for 44 percent of fuel consumption.

Georgia has substantial hydroelectric potential, only 14 percent of which was in use in 1993 in a network of small hydroelectric stations. In 1993 all but eight of Georgia's seventy-two power stations were hydroelectric, but together they provided only half the republic's energy needs. In the early 1990s, Georgia's total consumption of electrical energy exceeded domestic generation by as much as 30 percent. Georgian planners see further hydroelectric development as the best domestic solution to the country's power shortage.

Agriculture

Georgia's climate and soil have made agriculture one of its most productive economic sectors; the 18 percent of Georgian land that is arable provided 32 percent of the republic's NMP in 1990. In the Soviet period, swampy areas in the west were drained and arid regions in the east were salvaged by a complex irrigation system, allowing Georgian agriculture to expand production tenfold between 1918 and 1980. Production was hindered in the Soviet period, however, by the misallocation of agricultural land (for example, the assignment of prime grain fields to tea cultivation) and excessive specialization. Georgia's emphasis on labor-intensive crops such as tea and grapes kept the rural work force at an unsatisfactory level of productivity. Some 25 percent of the Georgian work force was engaged in agriculture in 1990; 37 percent had been so engaged in 1970 (see table 4). In the spring of 1993, sowing of spring crops was reduced by one third on state land and by a substantial amount on private land as well because of fuel and equipment shortages. For the first half of 1993, overall agricultural production was 35 percent less than for the same period of 1992.

Land Redistribution

Until the land-privatization program that began in 1992, most Georgian farms were state-run collectives averaging 428 hectares in size. Even under Soviet rule, however, Georgia had a vigorous private agricultural sector. In 1990, according to official statistics, the private sector contributed 46 percent of gross agricultural output, and private productivity averaged about twice that of the state farms (see table 5). Under the state system, designated plots were leased to farmers and town dwellers for private crop and livestock raising. As during the Soviet era, more than half of Georgia's meat and milk and nearly half of its eggs come from private producers.

Table 4. Georgia: Employment by Economic Activity, 1989, 1990, and 1991 (in thousands of people)

Activity	1989	1990	1991
Agriculture	656	695	671
Forestry	12	12	11
Industry	537	560	488
Construction	266	281	225
Transportation and communications	123	115	104
Trade and commercial services	267	257	227
Housing and municipal services	123	131	110
Science and research and development	73	73	63
Education and culture	301	310	290
Health, social welfare, and sports	189	184	186
Banking and financial	12	12	12
Government	55	52	48
Other services	86	82	79
TOTAL*	2,700	2,763	2,514

* Figures may not add to totals because of rounding.
Source: Based on information from *The Europa World Year Book, 1993,* 1, London, 1993, 1235.

As was the case with enterprise privatization, Gamsakhurdia postponed systematic land reform because he feared that local mafias would dominate the redistribution process. But within weeks of his ouster in early 1992, the new government issued a land reform resolution providing land grants of one-half hectare to individuals with the stipulation that the land be farmed. Commissions were established in each village to inventory land parcels and identify those to be privatized. Limitations were placed on what the new "owners" could do with their land, and would-be private farmers faced serious problems in obtaining seeds, fertilizer, and equipment. By the end of 1993, over half the cultivated land was in private hands. Small plots were given free to city dwellers to relieve the acute food shortage that year.

Table 5. Georgia: Population and Employment, 1988-91
(in thousands of people)

	1988	1989	1990	1991
Total population	5,396	5,414	5,422	5,421
Males	2,561	2,571	2,579	n .a.
Females	2,835	2,843	2,843	n.a.
Urban	2,989	3,014	3,029	3 ,024
Rural	2,407	2,400	2,393	2 ,397
Total employed	2,650	2,635	2,685	2,543
Males employed	n.a.	1,414	n.a.	n.a.
Females employed	n.a.	1,221	n.a.	n.a.
In state sector	2,205	2,148	2,087	1,886
In collective farms	249	218	200	154
Self-employed	177	211	229	356
Other	19	58	169	147

n.a.--not available.
Source: Based on information from World Bank, *Georgia: A Blueprint for Reforms,* Washington, 1993, 109-10.

Crop Distribution

In 1993 about 85 percent of cultivated land, excluding orchards, vineyards, and tea plantations, was dedicated to grains. Within that category, corn grew on 40 percent of the land, and winter wheat on 37 percent. The second most important agricultural product is wine. Georgia has one of the world's oldest and finest winemaking traditions; archeological findings indicate that wine was being made in Georgia as early as 300 B.C. Some forty major wineries were operating in 1990, and about 500 types of local wines are made. The center of the wine industry is Kakhetia in eastern Georgia. Georgia is also known for the high quality of its mineral waters.

Other important crops are tea, citrus fruits, and noncitrus fruits, which account for 18.3 percent, 7.7 percent, and 8.4 percent of Georgia's agricultural output, respectively. Cultivation of tea and citrus fruit is confined to the western coastal area. Tea accounts for 36 percent of the output of the large food-processing industry, although the quality of Georgian tea dropped perceptibly under Soviet management in the 1970s and 1980s. Animal husbandry, mainly the keeping of cattle, pigs, and sheep, accounts for about 25 percent of Georgia's agricultural output, although high density and low mechanization have hindered efficiency.

Until 1991 other Soviet republics bought 95 percent of Georgia's processed tea, 62 percent of its wine, and 70 percent of its canned goods (see table 6; table 7). In turn, Georgia depended on Russia for 75 percent of its grain. One-third of Georgia's meat and 60 percent of its dairy products were supplied from outside the republic. Failure to adjust these relationships contributed to Georgia's food crises in the early 1990s.

Table 6. Georgia: Annual Per Capita Food Consumption, Selected Years, 1970-90 (in kilograms unless otherwise specified)

Food	1970	1980	1985	1990
Bread	195	190	190	183
Eggs (in units)	85	135	148	140
Fish	6	8	9	9
Meat	31	43	47	46
Milk and dairy products	235	309	309	289
Potatoes	38	46	49	41
Sugar	35	45	43	39
Vegetable oil	3	5	6	6
Vegetables	51	79	87	82

Source: Based on information from World Bank, *Food and Agricultural Policy Reforms in the Former USSR*, Washington, 1992, 183-84.

Transportation and Telecommunications

Georgia's location makes it an important commercial transit route, and the country inherited a well-developed transportation system when it became independent in 1991. However, lack of money and political unrest have cut into the system's maintenance and allowed it to deteriorate somewhat since independence. Fighting in and around the secessionist Abkhazian Autonomous Republic in the northwest has isolated that area and also has cut some of the principal rail and highway links between Georgia and Russia.

In 1990 Georgia had 35,100 kilometers of roads, 31,200 kilometers of which were paved. Since the nineteenth century, Tbilisi has been the center of the Caucasus region's highway system, a position reinforced during the Soviet era. The country's four principal highways radiate from Tbilisi roughly in the four cardinal directions. Route M27 extends west from the capital through the broad valley between the country's two main mountain ranges and reaches the Black Sea south of Sukhumi. The highway then turns northwest along the Black Sea to the Russian border. A secondary road, Route A305, branches off Route M27 and carries traffic to the port of Poti. Another secondary road runs south along the Black Sea coast from Poti to the port of Batumi. From Batumi a short spur of about ten kilometers is Georgia's only paved connection with Turkey.

Route A301, more commonly known as the Georgian Military Highway, runs north for almost 200 kilometers from Tbilisi across the Greater Caucasus range to Russia. The route was first described by Greek geographers in the first century B.C. and was the only land route north into Russia until the late 1800s. The route contains many hairpin turns and winds through several passes higher than 2,000 meters in elevation before reaching the Russian border. Heavy snows in winter often close the road for short periods. The country's other two main highways connect Tbilisi with the neighboring Transcaucasian countries. Route A310 runs south to Erevan, and Route A302 extends east across a lower portion of the Greater Caucasus range to Azerbaijan. All major routes have regular and frequent bus transport.

Georgia had 1,421 kilometers of rail lines in 1993, excluding several small industrial lines. In the early 1990s, most lines were 1.520-meter broad gauge, and the principal routes were electrified. The tsarist government built the first rail links in the region from Baku on the Caspian Sea through Tbilisi to Poti on the Black Sea in 1883; this route remains the principal rail route of Transcaucasia. Along the Black Sea, a rail route extends from the main east-west line into Russia, and two lines run south from Tbilisi--one to Armenia and the other to Azerbaijan. Spurs link these main routes with smaller towns in Georgia's broad central valley. Principal classification yards and rail repair services are in Batumi and Tbilisi. Most rail lines provide passenger service, but in 1994 international passenger service was limited to the Tbilisi-to-Baku train. Because of fighting in Abkhazia, freight and passenger service to Russia has been suspended, with only the section from Tbilisi to the port of Poti still operative. Service on the Tbilisi-to-Erevan line has also been disrupted because the tracks pass through the area of armed conflict between Armenia and Azerbaijan.

Tbilisi was one of the first cities of the Soviet Union to have a subway system. The system consists of twenty-three kilometers of heavy rail lines, most of which are underground. Three lines with twenty stations radiate from downtown, with extensions either planned or under construction in 1994. The system is heavily used, and trains run at least every four minutes throughout the day. In 1985, the last year of available statistics, 145 million passengers were carried, about the same number of passengers that used Washington's Metrorail system in 1992.

Georgia's principal airport, Novoalekseyevka, is about eighteen kilometers northeast of downtown Tbilisi. With a runway approximately 2,500 meters long, the airport can accommodate airplanes as large as the Russian Tu-154, the Boeing 727, and the McDonnell Douglas DC-9. In 1993 the airport handled about 26,000 tons of freight. Orbis, the new state-run airline, provides service to neighboring countries, flights to several destinations throughout Russia, and direct service to some European capitals. Between 1991 and 1993, fuel shortages severely curtailed air passenger and cargo service, however. Eighteen other airports throughout the country have paved runways, but most are used for minor freight transport.

Georgia's Black Sea ports provide access to the Mediterranean Sea via the Bosporus. Georgia has two principal ports, at Poti and Batumi, and a minor port at Sukhumi. Although Batumi has a natural harbor, Poti's man-made harbor carries more cargo because of that city's rail links to Tbilisi. The port at Poti can handle ships having up to ten meters draught and 30,000 tons in weight. Altogether, nine berths can process as much as 100,000 tons of general cargo, 4 million tons of bulk cargo, and 1 million tons of grain per year. Facilities include tugboats, equipment for unloading tankers, a grain elevator, 22,000 square meters of covered storage area, and 57,000 square meters of open storage area. Direct onloading of containers to rail cars is available. The port primarily handles exports of grain, coal, and ores and imports of general cargo. Poti is ice-free, but in winter strong west winds can make entry into the port hazardous.

Batumi's natural port is located on a bay just northeast of the city. Eight alongside berths have a total capacity of 100,000 tons of general cargo, 800,000 tons of bulk cargo, and 6 million tons of petroleum products. Facilities include portal cranes, loaders for moving containers onto rail cars, 5,400 square meters of covered storage, and 13,700 square meters of open storage. The port lies at the end of the Transcaucasian pipeline from Baku and is used primarily for the export of petroleum and petroleum products. The port's location provides some protection from the winds that buffet Poti. However, strong winds can cause dangerous currents in the port area, forcing ships to remain offshore until conditions improve.

Sukhumi, capital of the Abkhazian Autonomous Republic, is a small port that handles limited amounts of cargo, passenger ferries, and cruise ships. Imports consist mostly of building materials, and the port handles exports of local agricultural products, mostly fruit. Strong westerly and southwesterly winds make the port virtually unusable for long periods in the autumn and winter. Sukhumi has been unavailable to Georgia since Georgian forces abandoned the city during the conflict of the autumn of 1993.

In 1992 Georgia had 370 kilometers of crude oil pipeline, 300 kilometers of pipeline for refined petroleum products, and 440 kilometers of natural gas pipeline. Batumi is the terminus of a major oil pipeline that transports petroleum from Baku across the Caucasus for export. Two natural gas pipelines roughly parallel the route of the oil pipeline from Baku to Tbilisi before veering north along the Georgian Military Highway to Russia. Pipelines are generally high-capacity lines and have a diameter of either 1,020 or 1,220 millimeters.

Historically, Georgia was an important point on the Silk Road linking China with Europe. Since independence Georgians have discussed resuming this role by turning the republic into a modern transportation and communications hub. Such a plan might also make the republic a "dry Suez" for the transshipment of Iranian oil west across the Caucasus.

In 1991 about 672,000 telephone lines were in use, providing twelve lines per 100 persons. The waiting list for telephone installation was quite long in the early 1990s. Georgia is linked to the CIS countries and Turkey by overland lines, and one low-capacity satellite earth station is in operation. Three television stations, including the independent Iberia Television, and numerous radio stations broadcast in Georgian and Russian.

Economic Reform

Like all the former Soviet republics, Georgia recognized the need to restructure its economic system in the early 1990s, using national economic strengths to accommodate its own needs rather than the needs of central planners in Moscow. The road to reform has been full of obstacles, however: poor political leadership, the economic decline that began in the 1980s, civil war, and a well-established underground economy that is difficult to control.

Price Policy

Gamsakhurdia understood little about economics, and he postponed major economic reforms to avoid weakening his political position. In an effort to maintain popular support, he stabilized fares for public transportation and prices for basic consumer goods in state retail outlets (see table 7). In March 1991, a new rationing system bound local residents to neighborhood shops. In April 1991, price controls were imposed in state stores. Price liberalization began only after Gamsakhurdia's departure as president, and it did not cover several basic consumer goods and services. Continued food subsidies were an additional factor contributing to the national budget deficit. In the interest of stimulating competition, a government decree removed restrictions on trade in May 1992, and at the same time taxes were eliminated on goods brought into Georgia. Persistent shortages of bread led the government to introduce ration cards for bread in December 1992. Under these conditions, inflation soared in private markets in 1991-92, although prices remained substantially lower than in Moscow for similar items.

Table 7. Georgia: Durable Consumer Goods, 1989, 1990, and 1991
(items per 100 families)

Product	1988	1989	1990
Automobiles	31	31	34
Refrigerators	95	95	95
Sewing machines	63	63	61
Tape recorders	44	48	52
Televisions	102	106	112
Washing machines	76	81	86

Source: Based on information from United States, Central Intelligence Agency, *Handbook of International Economic Statistics, 1993,* Washington, 1993, 77.

In 1993 wholesale prices increased especially quickly under the influence of falling productivity. In the second half of 1993, the construction industry was hit hard by increases in the cost of materials of up to thirty times, although gasoline prices rose only gradually. The prices of heavy engineering and ferrous-metallurgy products rose by three to five times in the second half of 1993.

Table 8. Georgia: External Trade, 1988, 1989, and 1990
(in millions of rubles)

	1988	1989	1990
Exports			
Oil and gas	100	68	68
Ferrous metallurgy	375	376	318
Chemical fuels	316	343	339
Machines and processed metals	848	869	804
Nonfood light industrial products	1,275	1,285	1,260
Processed foods	2,438	2,573	2,387
Other industrial products	258	275	310
Agricultural products	280	190	404
Other	11	105	93
Total exports	5,901	6,084	5,983
To other Soviet republics	5,508	5,719*	5,724
To other countries	393	465*	259
Imports			
Oil and gas	413	360	285
Ferrous metallurgy	489	443	430
Nonferrous metallurgy	102	106	97
Chemical fuels	541	544	576
Machines and processed metals	1,533	1,522	1,580
Timber and wood products	248	244	279
Building materials	155	148	117
Nonfood light industrial products	1,221	1,287	1,372
Processed foods	1,204	1,142	1,174
Other industrial products	212	212	291
Agricultural products	348	358	498
Other	27	103	140
Total imports	6,493	6,469	6,839
From other Soviet republics	5,218	4,888	4,948
From other countries	1,275	1,581	1,891

* As published.

Source: Based on information from *The Europa World Year Book, 1994,* 1, London, 1994, 1237.

Enterprise Privatization

Another key element of economic reform, privatization of state enterprises, was stifled under Gamsakhurdia. He feared that the "economic mafia," which already owned a significant share of the nation's wealth, would use that wealth to accumulate state assets. Rapid growth had already occurred in the private retail sector, however, once cooperative enterprises began expanding in 1988. In 1990-91 privately run "commercial shops" began proliferating, often in place of state stores. Typically, these shops offered consumer goods brought from Turkey and resold at very high prices. The Law on Privatization of State Enterprises was adopted in August 1991 to outline general principles, and the Committee on Privatization was established in 1992. Under Shevardnadze, privatization began cautiously in August 1992 when the State Council adopted the State Program on the Privatization of State Enterprises. The law copied Russia's approach to privatization by providing for several methods, including "popular privatization," consisting of a combination of vouchers distributed to the public and auctions of state enterprises. The country's political crises delayed meaningful measures, however. By 1993 few Georgian industries had been privatized, although large numbers of small enterprises were scheduled for privatization in 1993 and 1994.

Foreign Trade

In the Soviet period, Georgian trade with the world outside the Soviet Union was severely restricted by Moscow's foreign economic policy (see table 8). Almost all of Georgian foreign economic activity was conducted by fourteen central enterprises, most of which operated under the direct management of Moscow. Bulgaria, Czechoslovakia, Germany, Japan, and Poland were among the most important of Georgia's trading partners (see table 9). Gamsakhurdia, suspicious of businessmen who sought to export Georgian goods, banned all export activity. The Shevardnadze government, however, created conditions for significant improvement of international investment and trade. In May 1992, licensing requirements for import or export activities were dropped except for the import of goods in the military and medical categories. This change represented a significant expansion of the rights of enterprises to engage in foreign economic activity. Export of twelve commodities, mostly foodstuffs, was still prohibited at the end of 1992. Fees and other restrictions on the registration of joint ventures were removed, and the state tax on all imports was canceled. Import duties ranged from 5 to 55 percent, and export duties from 5 to 90 percent, with an exemption for former Soviet republics; the VAT on exports dropped to 14 percent in late

1992. The National Bank of Georgia imposed a tax of 12 percent of exporters' hard-currency earnings. In early 1993, new trade policies had not led to major increases in foreign trade and investment. Continued political instability, ethnic warfare, and extremely poor transportation and telecommunications facilities continued to discourage foreign investors in 1993.

In the second half of 1993, continued military upheaval did not entirely deter progress in foreign investment. The Renault automobile company of France, the German Tee Kanes tea company, and British and Dutch liquor companies signed contracts in August, and officials of Mitsubishi and an American shipbuilder visited Georgia to assess investment conditions.

Table 9. Georgia: Major Trading Partners, 1990

Country	Value*		Percentage of Total	
	Exports	Imports	Exports	Imports
Europe	213	1,687	71.4	71.2
Austria	6	53	2.0	2.2
Bulgaria	32	202	10.7	8.5<
ungary	18	144	6.0	6.1
Italy	13	53	4.4	2.2
Poland	25	256	8.4	10.8
Germany	32	420	10.7	17.7<
inland	7	76	2.3	3.2
Czechoslovakia	23	227	7.7	9.6
Yugoslavia	8	64	2.7	2.7
Asia	45	377	15.2	15.9
India	6	48	2.0	2.0
China	6	34	1.9	2.7
Syria	1	76	0.2	3.2
Japan	12	115	4.0	4.9
North America	22	192	7.3	8.1
Cuba	16	114	5.4	4.8
United States	3	37	1.0	1.6

* In millions of rubles.
Source: Based on information from World Bank, *Statistical Handbook: States of the Former USSR,* Washington, 1992, 152.

GOVERNMENT AND POLITICS

In the late 1980s and the early 1990s, the tone of Georgian political life changed significantly. National elections held in 1989, 1990, and 1992 reflected that change. The nature of governance in newly independent Georgia was most influenced by the personalities of two men, Zviad Gamsakhurdia and Eduard Shevardnadze. But democratic institutions evolved slowly and sporadically in the first half of the 1990s.

Communist-built secular wedding chapel, Tbilisi
Courtesy Courtesy Michael W. Serafin

Establishing Democratic Institutions

Prior to the 1989 elections, the Georgian Communist Party maintained tight control over the nomination process. Even in 1989, candidates ran unopposed in forty-three of seventy-five races, and elsewhere pairings with opposition candidates were manipulated to guarantee results favoring the party. In Tbilisi grassroots movements succeeded in nominating three candidates to the Georgian Supreme Soviet in 1989. The leaders of these movements were mostly young intellectuals who had not been active dissidents. Many of those figures later joined to form a new political party, Democratic Choice for Georgia, abbreviated as DASi in Georgian. Because of expertise in local political organization, DASi played a leading role in drafting legislation for local and national elections between 1990 and 1992.

The death of the Tbilisi demonstrators in April 1989 led to a major change in the Georgian political atmosphere. Radical nationalists such as Gamsakhurdia were the primary beneficiaries of the national outrage following the April Tragedy. In his role as opposition leader, Gamsakhurdia formed a new political bloc in 1990, the Round Table/Free Georgia coalition.

In 1990 Georgia was the last Soviet republic to hold elections for the republic parliament. Protests and strikes against the election law and the nominating process had led to a six-month postponement of the elections until October 1990. Opposition forces feared that the political realities favored entrenched communist party functionaries and the enterprise and collective farm officials they had put in place. According to reports, about one-third of the 2,300 candidates for the Supreme Soviet (as the Georgian parliament was still designated at that time) fell into this category.

The electoral system adopted in August 1990, which represented a compromise between competing versions put forward by the Patiashvili government and the opposition, created the first truly multiparty elections in the Soviet Union. The new Georgian election law combined district-level, single-mandate, majority elections with a proportional party list system for the republic as a whole; a total of 250 seats would constitute the new parliament. On one hand, the proportional voting system required that a party gain at least 4 percent of the total votes to achieve representation in parliament. On the other hand, candidates with strong local support could win office even if their national totals fell below the 4 percent threshold. When the elections finally were held, widespread fears of violence or communist manipulation (expressed most vocally by Gamsakhurdia) proved unfounded.

The 1990 Election

The 1990 parliamentary election was a struggle between what remained of the Georgian Communist Party, which still held power at that point, and thirty-one opposition parties constituting the Georgian national movement. The national movement was not completely represented in the official election, however, because many opposition parties organized separate elections to an alternative body called the Georgian National Congress. An important factor in the results was a provision in the election law that forbade members of the communist party to run simultaneously on the ticket of another party. (By contrast, in this interim period other Soviet republics allowed even proponents of radical reform to retain their communist party memberships while representing popular fronts and similar organizations.)

The election decisively rejected the communists and gave a resounding popular mandate to the Round Table/Free Georgia bloc that Gamsakhurdia headed. That coalition captured 54 percent of the proportional vote to gain 155 seats out of the 250 up for election, while the communists gained 64 seats and 30 percent of the proportional vote. Communist strongholds remained in Azerbaijani and Armenian districts of southern Georgia. No other party reached the 4 percent share necessary for representation in the party-list system, and only a handful of candidates from other parties won victories in the individual district races. Boycotts prevented voting in two districts of Abkhazia and in two districts of South Ossetia.

Gamsakhurdia raised initial hopes for compromise in his new government by withdrawing Round Table/Free Georgia candidacies from runoffs against the opposition Popular Front Party in twelve races. That move ensured the election of Popular Front candidates as individuals in those contests; otherwise, the 4 percent rule would have precluded representation for the Popular Front.

The Gamsakhurdia Government

Gamsakhurdia's choice to head the new government, Tengiz Sigua, was almost universally praised. Sigua, formerly director of a metallurgy institute, had been an adroit and evenhanded deputy chairman of the Central Election Commission supervising the 1990 election. The government formed by Gamsakhurdia included many officials who lacked previous government experience. Only one full minister was retained from the communist government, although former deputy ministers were frequently promoted to the top post in ministries concerned with the economy. Initially, the large number of remaining communist deputies formed no organized opposition bloc in the parliament. In fact, the communist party faded rapidly from the scene, and most of its property and publishing facilities were seized. The large, modern facility Shevardnadze had built for the party's Central Committee was taken over by the Cabinet of

Ministers. The rapid decline of the communists showed that the major attraction of communist party membership had been the party's position of power; once that power was lost, the number of active communists dropped almost to zero. When the new first secretary of the party ran against Gamsakhurdia for president in 1991, he received less than 2 percent of the vote. After the August 1991 coup in Moscow, Gamsakhurdia banned the communist party, and deputies elected to parliament on the communist ticket were deprived of their seats.

Gamsakhurdia's Ouster and Its Aftermath

A small but vocal parliamentary opposition to Gamsakhurdia began to coalesce after August 1991, particularly after government forces reportedly fired on demonstrators in September. At this time, several of Gamsakhurdia's top supporters in the Round Table/Free Georgia bloc joined forces with the opposition. However, the opposition was unable to convince Gamsakhurdia to call new elections in late 1991. The majority of deputies, most of whom owed their presence in parliament to Gamsakhurdia, supported him to the end. Indeed, a significant number of deputies followed Gamsakhurdia into exile in Chechnya, where they continued to issue resolutions and decrees condemning the "illegal putsch."

In the aftermath of Gamsakhurdia's ouster in January 1992, parliament ceased to function and an interim Political Consultative Council was formed. It was to consist of about forty members, to include ten political parties, a select group of intellectuals, and several opposition members of parliament. This council was intended to serve as a substitute parliament, although it only had the right to make recommendations. Legislative functions were granted to a new and larger body, the State Council, created in early March 1992. By May 1992, the State Council had sixty-eight members, including representatives of more than thirty political parties and twenty social movements that had opposed Gamsakhurdia. Efforts were also made to bring in representatives of Georgia's ethnic minorities, although no Abkhazian or Ossetian representatives participated in the new council.

Almost immediately after Gamsakhurdia's ouster, Sigua resumed his position as prime minister and created a working group to draft a new election law that would legitimize the next elected government. Immediately after the overthrow of Gamsakhurdia, the new government feared that Gamsakhurdia retained enough support in Georgia to regain power in the next election. As a result, in March the State Council adopted an electoral system, the single transferable vote, which would virtually guarantee representation by small parties and make it difficult for a party list headed by one prominent figure to translate a majority of popular votes into parliamentary control.

New Parties and Shevardnadze's Return

After his return to Georgia in March 1992, Shevardnadze constantly stressed the temporary nature of the new power structure and called for elections as soon as possible. But the leadership postponed balloting until October 1992 because it lacked effective political control over many regions of the country and because of factional wrangling over the new election law. Registration of political parties, which had been suspended by Gamsakhurdia in 1991, resumed early in 1992. Among new party registrants was the Democratic Union, a group consisting mostly of former members and officials of the communist party. Claiming a broad mass following, this party had organizations in most regions of the county. Although wooed by the Democratic Union and other parties, Shevardnadze avoided party affiliation in order to maintain his independent position. The parliament that would be elected in October 1992 clearly would be an interim body given the task of writing a new constitution. Accordingly, the term of office was set for three years.

The Election of 1992

After a series of last-minute changes, the electoral system for October 1992 was a compromise combination of single-member districts and proportional voting by party lists. To give regional parties a chance to gain representation, separate party lists were submitted for each of ten historical regions of Georgia. In a change from the 1990 system, no minimum percentage was set for a party to achieve representation in parliament if the party did sufficiently well regionally to seat candidates. Forty-seven parties and four coalitions registered to participate in the 1992 election. For the first time, the Central Election Commission accepted the registration of every party that submitted an application.

Burned-out headquarters building of Georgian Communist Party, Tbilisi, 1992
Courtesy Michael W. Serafin

The largest of the electoral alliances, and one of the most controversial, was the Peace Bloc (Mshvidoba). This broad coalition of seven parties ranged from the heavily ex-communist Democratic Union to the Union for the Revival of Ajaria, a party of the conservative Ajarian political elite. Ultimately, the strong programmatic differences among the seven parties would render the Peace Bloc ineffective as a parliamentary faction. The Democratic Union filled as much as 70 percent of the places given the coalition on the party lists. In the 1992 election, the Peace Bloc draw a plurality of votes, thus earning the coalition twenty-nine seats in parliament.

The second most important coalition, the October 11 Bloc, included moderate reform leaders of four parties. Members typically had academic backgrounds with few or no communist connections, and the median age of bloc leaders was about fifteen years less than that of the Democratic Union leadership. The October 11 Bloc won eighteen seats, the second largest number in the 1992 election.

A third coalition, the Unity Bloc (Ertoba), lost two of its four member parties before the election. Many of the leaders of the Liberal-Democratic National Party, one of the two remaining constituent parties of the Unity Bloc, were, like the leaders of the Democratic Union, former communist officials who continued to hold influential posts in the Georgian government and mass media. Both the Peace Bloc and the Unity Bloc put prominent cultural figures at the top of their electoral lists to gain attention.

Shevardnadze's actions were crucial in building the foundation for the 1992 election. From the time of his return to Georgia, Shevardnadze enjoyed unparalleled respect and recognition. Because of his unique position, the State Council acted to separate Shevardnadze from party politics by creating a potentially powerful new elected post, chairman of parliament, which would also be contested in the October elections. Because no other candidate emerged, Shevardnadze was convinced to forego partisan politics and grasp this opportunity for national leadership.

The elections took place as scheduled in October 1992 in most regions of the country. International monitors from ten nations reported that, with minor exceptions, the balloting was free and fair. Predictably, Gamsakhurdia declared the results rigged and invalid. Interethnic tensions and Gamsakhurdia's activity forced postponement of elections in nine of the eighty-four administrative districts, located in Abkhazia, South Ossetia, and western Georgia. Voters in those areas were encouraged to travel to adjoining districts, however, to vote in all but the regional races. Together, the nonvoting districts represented 9.1 percent of the registered voters in Georgia. In no voting district did less than 60 percent of eligible voters participate.

An important factor in the high voter turnout was the special ballot for Shevardnadze as chairman of the new parliament; a large number of voters cast ballots only for Shevardnadze and submitted blank or otherwise invalid ballots for the other races. Shevardnadze received an overwhelming endorsement, winning approximately 96 percent of the vote. In all, fifty-one of the ninety-two members of the previous State Council were elected to the new parliament. The four sitting members of the State Council Presidium (Shevardnadze, Ioseliani, Sigua, and Kitovani) also were reelected.

Formation of the Shevardnadze Government

An immediate goal after Shevardnadze's return was to avoid repeating the one-man rule imposed by Gamsakhurdia while keeping a sufficiently tight grip on central power to prevent regional separatism. The newly elected parliament convened for the first time in November 1992. The lack of dominant parties and the large number of independent deputies ensured that Shevardnadze would dominate parliamentary sessions. The precise role of Shevardnadze was not clear at the time of the elections; on November 6, the parliament ratified proposals on this subject in the Law on State Power. Instead of reestablishing the post of president that had been created by--and was still claimed by--Gamsakhurdia, parliament gave Shevardnadze a new title, head of government. In theory, parliament was to elect the holder of this office, although in practice the position was understood to be combined with the popularly elected post of chairman of the parliament. Thus an impasse between the executive and the legislative branches was avoided by giving the same person a top role in both, but the division of power between the branches remained unclear in early 1994.

The Cabinet

The government team selected by Shevardnadze, called the Cabinet of Ministers, was quickly approved by parliament in November 1992. Sigua returned as prime minister. Four deputy prime ministers were chosen in November 1992, including Tengiz Kitovani, former head of the National Guard and minister of defense in the new cabinet. In December 1992, the Presidium of the Cabinet of Ministers was created. This body included the prime minister and his deputy prime ministers as well as the minister of agriculture, the minister of finance, the minister of state property management, the minister of economics, and the minister of foreign affairs.

In December 1992, the Georgian government included eighteen ministries, four state committees, and fifteen departments, which together employed more than 7,600 officials. Many appointees to top government posts, including several ministers, had held positions in the apparatus of the Georgian Communist Party. Although Shevardnadze's early appointments favored his contemporaries and former associates, by late 1993 about half of the top state administrative apparatus were academics. Less than 10 percent were former communists, about 75 percent were under age forty, and more than half came from opposition parties.

In September 1993, the cabinet included the following ministries: agriculture and the food industry; communications; culture; defense; economic reform; education; environment; finance; foreign affairs; health; industry; internal affairs; justice; labor and social security; state property management; and trade and supply. Each of the five deputy prime ministers supervised a group of ministries.

In practice, the Cabinet of Ministers was a major obstacle to reform in 1993. Pro-reform ministers were isolated by the domination of former communists in the Presidium, which stood between Shevardnadze and the administrative machinery of the ministries. In 1993 Shevardnadze himself was reluctant to push hard for the rapid reforms advocated by progressives in parliament. The cabinet was superficially restructured in August 1993, but reformers clamored for a smaller cabinet under direct control of the head of state.

Parliament

In 1993 some twenty-six parties and eleven factions held seats in the new parliament, which continued to be called the Supreme Soviet. The legislative branch's basic powers were outlined in the Law on State Power, an interim law rescinding the strict limits placed on legislative activity by Gamsakhurdia's 1991 constitution. Thus in 1993 the parliament had the power to elect and dismiss the head of state by a two-thirds vote; to nullify laws passed by local or national bodies if they conflicted with national law; to decide questions of war and peace; to reject any candidate for national office proposed by the head of state; and, upon demand of one-fifth of the deputies, to declare a vote of no confidence in the sitting cabinet.

Activity within the legislative body was prescribed by the Temporary Regulation of the Georgian Parliament. The parliament as a whole elected all administrative officials, including a speaker and two deputy speakers. Seventeen specialized commissions examined all bills in their respective fields. The speaker had little power over commission chairs or over deputies in general, and parliament suffered from an inefficient structure, insufficient staff, and poor communications. The two days per week allotted for legislative debate often did not allow full consideration of bills.

The major parliamentary reform factions--the Democrats, the Greens, the Liberals, the National Democrats, and the Republicans--were not able to maintain a coalition to promote reform legislation. Of that group, the National Democrats showed the most internal discipline. Shevardnadze received support from a large group of deputies from single-member districts, aligned with Liberals and Democrats. His radical opposition, a combination of several very small parties, was weakened by disunity, but it frequently was able to obstruct debate. The often disorderly parliamentary debates reduced support among the Georgian public, to whom sessions were widely televised.

In November 1993, Shevardnadze was able to merge three small parties with a breakaway faction of the Republicans to form a new party, the Union of Citizens of Georgia, of which he became chairman. This was a new step for the head of state, who previously had refrained from political identification and had relied on coalitions to support his policies. At the same time, Shevardnadze also sought to include the entire loose parliamentary coalition that had recently supported him, in a concerted effort to normalize government after the Abkhazian crisis abated.

The Chief Executive

The 1992 Law on State Power gave Shevardnadze power beyond the executive functions of presidential office. As chairman of parliament, he had the right to call routine or extraordinary parliamentary sessions, preside over parliamentary deliberations, and propose constitutional changes and legislation. As head of state, Shevardnadze nominated the prime minister, the cabinet, the chairman of the Information and Intelligence Service, and the president of the National Bank of Georgia (although the parliament had the right of approval of these officials).

Without parliamentary approval, the head of government appointed all senior military leaders and provincial officials such as prefects and mayors. Additional power came from his control of the entire system of state administration, and he could form his own administrative apparatus, which had the potential to act as a shadow government beyond the control of any other branch. Key agencies chaired by Shevardnadze in 1993 were the Council for National Security and Defense, the Emergency Economic Council, and the Scientific and Technical Commission, which advised on military and industrial questions.

In response to calls by the opposition for his resignation during the Abkhazian crisis of mid-1993, Shevardnadze requested and received from parliament emergency powers to appoint all ministers except the prime minister and to issue decrees on economic policy without legislative approval. When the Sigua government resigned in August, parliament quickly approved Shevardnadze's nomination of industrialist Otar Patsatsia as prime minister. Although Shevardnadze argued that greater central power was necessary to curb turmoil, his critics saw him setting a precedent for future dictatorship and human rights abuses.

The Judicial System

When Georgia was part of the Soviet Union, the Supreme Court of Georgia was subordinate to the Supreme Court of the Soviet Union, and the rule of law in Georgia, still based largely on the Soviet constitution, included the same limitations on personal rights. Beginning in 1990, the court system of Georgia began a major transition toward establishment of an independent judiciary that would replace the powerless rubber-stamp courts of the Soviet period. The first steps, taken in late 1990, were to forbid Supreme Court judges from holding communist party membership and to remove Supreme Court activities from the supervision of the party. After the overthrow of Gamsakhurdia, the pre-Soviet constitution of 1921 was restored, providing the legal basis for separation of powers and an independent court. Substantial opposition to actual independence was centered in the Cabinet of Ministers, however, some of whose members would lose de facto judicial power.

The Supreme Court

In 1993 the Supreme Court had thirty-nine members, of whom nine worked on civil cases and thirty on criminal cases. All judges had been elected for ten-year terms in 1990 and 1991. Shevardnadze made no effort to replace judges elected under Gamsakhurdia, although they had been seated under a different constitutional system. The Supreme Court's functions include interpreting laws, trying cases of serious criminal acts and appeals of regional court decisions, and supervising application of the law by other government agencies.

The Procurator General

The postcommunist judicial system has continued the multiple role of the procurator general's office as an agency of investigation, a constitutional court supervising the application of the law, and the institution behind prosecution of crimes in court. In 1993 the procurator general's office retained a semimilitary structure and total authority over the investigation of court cases; judges had no power to reject evidence gained improperly. Advocates of democratization identified abolition of the office of procurator general as essential, with separation of the responsibilities of the procurator general and the courts as a first step.

Prospects for Reform

All parties in Georgia agreed that judicial reform depended on passage of a new constitution delineating the separation of powers. If such a constitution prescribed a strong executive system, the head of government would appoint Supreme Court judges; if a parliamentary system were called for, parliament would make the court appointments. In early 1994, however, the constitution was the subject of prolonged political wrangling that showed no sign of abating. At that point, experts found a second fundamental obstacle to judicial reform in a national psychology that had no experience with democratic institutions and felt most secure with a unitary, identifiable government power. Reform was also required in the training of lawyers and judges, who under the old system entered the profession through the sponsorship of political figures rather than on their own merit.

Regional Courts

Until the Gamsakhurdia period, regional courts were elected by regional party soviets; since 1990 regional courts have been appointed by regional officials. After the beginning of ethnic struggles in South Ossetia and Abkhazia, regional military courts also were established. The head of state appoints military judges, and the Supreme Court reviews military court decisions. The Tbilisi City Court has separate jurisdiction in supervising the observance of laws in the capital city.

The Constitution

Under Gamsakhurdia Georgia had continued to function under the Soviet-era constitution of 1978, which was based on the 1977 constitution of the Soviet Union. The first postcommunist parliament amended that document extensively. In February 1992, the Georgian National Congress (the alternate parliament elected in 1990) formally designated the Georgian constitution of February 21, 1921, as the effective constitution of Georgia. That declaration received legitimacy from the signatures of Jaba Ioseliani and Tengiz Kitovani, at that time two of the three members of the governing Military Council.

In February 1993, Shevardnadze called for extensive revisions of the 1921 constitution. Characterizing large sections of that document as wholly unacceptable, Shevardnadze proposed forming a constitutional commission to draft a new version by December 1993. According to Shevardnadze's timetable, the draft would be refined by parliament in the spring of 1994 and then submitted for approval by popular referendum in the fall of 1994.

Human Rights

Human rights protection and media freedom have been hindered in postcommunist Georgia by the national government's assumption of central executive power to deal with states of political and military emergency and by the existence of semi-independent military forces. In 1993 the expression of opposition views in the independent media was interrupted by official and unofficial actions against newspapers and broadcasters, despite a stated policy that expression of antigovernment views would be tolerated if not accompanied by violent acts.

Both sides of the Abkhazian conflict claimed widespread interference with civilian human rights by their opponents. Among the charges were abuse of military prisoners, the taking of civilian hostages, and the shelling and blockading of civilian areas. In 1993 the Shevardnadze government began addressing claims of human rights abuses by its military forces and police, particularly against Gamsakhurdia partisans and the Abkhazian population. In January the Parliamentary Commission on Human Rights and Ethnic Minority Affairs formed the Council of Ethnic Minorities, which met with representatives of the Meskhetian Turk exile population to resolve the grievances of that group. At the same time, the Interethnic Congress of the People of Georgia was formed to improve ethnic Georgians' appreciation of minority rights.

Despite the government's efforts, the Abkhazian conflict continued the tension between necessary wartime controls and the need to protect human rights. In June 1993, the international human rights group Helsinki Watch cited Georgia for political persecution, media obstruction, and military abuses of civilian rights, and in October the United States listed human rights progress as a prerequisite for continued economic aid.

The Media

The 1992 Law on the Press nominally reversed the rigorous state censorship of the Soviet and Gamsakhurdia periods and guaranteed freedom of speech. In 1993 Georgian law contained no prohibition of public criticism of the head of state, and Shevardnadze was subjected to accusations and comments from every direction. Three television channels are in operation; one, Ibervision, is run independently. Numerous independent newspapers are published; *Sakartvelos Respublika* (The Georgian Republic) presents the official government view in the daily press.

Despite some liberalization, in 1994 national security remained a rationale for media restriction. During the crisis of September 1993, two pro-Gamsakhurdia newspapers were closed and the office of an independent weekly were attacked by gunmen. The Free Media Association, an organization including eight independent newspapers, blamed a progovernment party for the attack. After his controversial decision in October to join the CIS, Shevardnadze threatened to close hostile newspapers, and no television channel discussed the widespread disagreement with the head of state's CIS initiative.

FOREIGN RELATIONS

Georgia's long tradition as a crossroads of East-West commerce was interrupted by the trade practices of the Soviet Union and then by Gamsakhurdia's isolationist policy. Although the Shevardnadze government sought to revive the national economy by reinstating ties with both East and West, in 1992 and 1993 domestic turmoil prevented major steps in that direction. In 1993 Shevardnadze traveled widely among the former Soviet republics (Azerbaijan, Kazakhstan, Russia, and Turkmenistan) and elsewhere (Germany, China, and the headquarters of the North Atlantic Treaty Organization in Belgium) to solidify Georgia's international position and solicit aid. By September, Georgia had diplomatic relations with seventy-eight countries and economic cooperation treaties with sixteen.

The Soviet and Gamsakhurdia Periods

Soviet policy effectively cut traditional commercial and diplomatic links to Turkey, which became a member of the North Atlantic Treaty Organization (NATO--see Glossary) in 1952, and to Iran, a United States ally until the late 1970s. Instead, virtually all transportation and commercial links were directed to Russia and the other Soviet republics. The same redirection occurred with diplomatic ties, which the Ministry of Foreign Affairs of the Soviet Union controlled. Shevardnadze's presence as Soviet foreign minister from 1985 to 1990 provided little direct benefit to Georgia aside from the large number of highranking guests who visited the republic in that period. That group included Britain's Prime Minister Margaret Thatcher and United States Secretary of State George Shultz.

Under Gamsakhurdia Georgia's efforts to break out of the diplomatic isolation of the Soviet period were stymied by the reluctance of the outside world to recognize breakaway republics while the Soviet Union still existed. Romania, which granted recognition in August 1991, was one of the few countries to do so during the Gamsakhurdia period. Several Georgian delegations came to the

United States in 1991 in an effort to establish diplomatic ties, but Washington largely ignored those efforts. Given stable internal conditions, the dissolution of the Soviet Union in late 1991 would have released Georgia from its isolation, but by that time the revolt against Gamsakhurdia was in full force. After the violent overthrow of Gamsakhurdia, other governments were reluctant to recognize the legitimacy of his successors. This situation changed in March 1992, when the internationally prominent Shevardnadze returned to Georgia and became chairman of the State Council.

In 1992 and 1993, United States aid to Georgia totaled US$224 million, most of it humanitarian, placing Georgia second in per capita United States aid among the former Soviet republics. In September 1993, Shevardnadze appealed directly to the United States Congress for additional aid. At that time, President William J. Clinton officially backed Shevardnadze's efforts to maintain the territorial integrity of Georgia. Reports of human rights offenses against opposition figures, however, brought United States warnings late in 1993 that continued support depended on the Georgian government's observance of international human rights principles.

The Foreign Policy Establishment

In his role as head of the State Council, Shevardnadze exerted a strong and direct influence on Georgia's foreign policy prior to the 1992 election. The additional post of head of government, which he acquired after the election, gave him the right to conduct negotiations with foreign governments and to sign international treaties and agreements. In the Sigua cabinet, the Ministry of Foreign Affairs was headed by Alexander Chikhvaidze, who had worked previously in the Ministry of Foreign Affairs of the Soviet Union and was serving as Soviet ambassador to the Netherlands at the time of his appointment in Tbilisi. The Council for National Security and Defense was formed in late 1992 to formulate strategic and security policy under the chairmanship of the head of state (see National Security).

Eduard Shevardnadze on official visit to the United States with President William J. Clinton, March 1994. Courtesy White House Photo Office

Revived Contacts in 1992

Shevardnadze's diplomatic contacts and personal relationships with many of the world's leaders ended Georgia's international isolation in 1992. In March Germany became the first Western country to post an ambassador to Georgia; Shevardnadze's close relations with German foreign minister Hans-Dietrich Genscher were a key factor in that decision. Recognition by the United States came in April 1992, and a United States embassy was opened in June 1992. Georgia became the 179th member of the United Nations in July 1992; it was the last of the former Soviet republics to be admitted. By December 1992, six countries had diplomatic missions in Tbilisi: China, Germany, Israel, Russia, Turkey, and the United States. Seventeen other countries began conducting diplomatic affairs with Georgia through their ambassadors to Russia or Ukraine. In August 1993, the United States granted Georgia most-favored-nation status, and the European Community offered technical economic assistance.

Unlike some former Soviet republics such as Armenia, Lithuania, and Ukraine, Georgia lacked a large number of emigrants in the West who could establish links to the outside world once internal conditions made such connections possible. Small groups of Georgian exiles lived in Paris and other European capitals, but they were mostly descended from members of the Social Democratic government that had been forced into exile with the incorporation of Georgia into the Soviet empire in 1921.

The only large group of emigrants that maintained contact with Georgia were Georgian Jews who had taken advantage of the Soviet Union's expansion of Jewish emigration rights in the 1970s and 1980s. Because Jews had lived in Georgia for many centuries and because Georgia had no history of anti-Semitism, many Georgian Jews continued to feel an attachment to Georgia and its culture, language, and people. Largely as a result of these ties, relations between Georgia and Israel flourished on many levels.

Relations with Neighboring Countries

Of particular importance to Georgia's postcommunist foreign policy and national security was the improvement of relations with neighbors on all sides: Armenia, Azerbaijan, Russia, and Turkey. This goal was complicated by a number of ethnic and political issues as well as by historical differences.

Armenia and Azerbaijan

Among former Soviet republics, the neighboring Transcaucasian nations of Armenia and Azerbaijan have special significance for Georgia. Despite Georgia's obvious cultural and religious affinities with Armenia, relations between Georgia and Muslim Azerbaijan generally have been closer than those with Christian Armenia. Economic and political factors have contributed to this situation. First, Georgian fuel needs make good relations with Azerbaijan vital to the health of the Georgian economy. Second, Georgians have sympathized with Azerbaijan's position in the conflict between Armenia and Azerbaijan over the ethnic Armenian enclave of Nagorno-Karabakh because of similarities to Georgia's internal problems with Abkhazia and South Ossetia. Both countries cite the principle of "inviolability of state borders" in defending national interests against claims by ethnic minorities (see figure 3; Nagorno-Karabakh and Independence; After Communist Rule).

In December 1990, Georgia under Gamsakhurdia signed a cooperation agreement with Azerbaijan affecting the economic, scientific, technical, and cultural spheres. In February 1993, Georgia under Shevardnadze concluded a far-reaching treaty of friendship, cooperation, and mutual relations with Azerbaijan, including a mutual security arrangement and assurances that Georgia would not reexport Azerbaijani oil or natural gas to Armenia. In 1993 Azerbaijan exerted some pressure on Georgia to join the blockade of Armenia and to curb incursions by Armenians from Georgian territory into Azerbaijan. The issue of discrimination against the Azerbaijani minority in Georgia, a serious matter during Gamsakhurdia's tenure, was partially resolved under Shevardnadze.

In the early 1990s, Armenia maintained fundamentally good relations with Georgia. The main incentive for this policy was the fact that Azerbaijan's blockade of Armenian transport routes and pipelines meant that routes through Georgia were Armenia's only direct connection with the outside world. Other considerations in the Armenian view were the need to protect the Armenians in Georgia and the need to stem the overflow of violence from Georgian territory. The official ties that Georgia forged with Azerbaijan between 1991 and 1993 strained relations with Armenia, which was in a state of virtual war with Azerbaijan for much of that period. Nevertheless, Gamsakhurdia signed a treaty with Armenia on principles of cooperation in July 1991, and Shevardnadze signed a friendship treaty with Armenia in May 1993. With the aim of restoring mutually beneficial economic relations in the Caucasus, Shevardnadze also attempted (without success) to mediate the Armenian-Azerbaijani conflict in early 1993.

Russia

Of all countries, Georgia's relations with Russia were both the most important and the most ambivalent. Russia (and previously the Soviet Union) was deeply involved at many levels in the conflicts in South Ossetia and Abkhazia, and in 1993 Ajarian leaders also declared Russia the protector of their national interests. Thus Russia seemingly holds the key to a resolution of those conflicts in a way that would avoid the fragmentation of Georgia. Trade ties with Russia, disrupted by Gamsakhurdia's struggle with Gorbachev and by ethnic conflicts on Georgia's borders with Russia, also are critical to reviving the Georgian economy.

Russia finally recognized Georgia's independence in mid-1992 and appointed an ambassador in October. In 1993 Russia's official position was that a stable, independent Georgia was necessary for security along Russia's southern border. The conditions behind that position were Russia's need for access to the Black Sea, which was endangered by shaky relations with Ukraine, the need for a buffer between Russia and Islamic extremist movements Russia feared in Turkey and Iran, the need to protect the 370,000 ethnic Russians in Georgia, and the refugee influx and violence in the Russian Caucasus caused by turmoil across the mountains in Georgia. Although Shevardnadze was officially well regarded,

Russian nationalists, many of them in the Russian army, wished to depose him as punishment for his initial refusal to bring Georgia into the CIS and for his role as the Soviet foreign minister who "lost" the former Soviet republics in 1991.

In pursuing its official goals, Russia offered mediation of Georgia's conflicts with the Abkhazian, Ajarian, and Ossetian minorities, encouraging Georgia to increase the autonomy of those groups for the sake of national stability. At the same time, Russian military policy makers openly declared Georgia's strategic importance to Russian national security. Such statements raised suspicions that, as in 1801 and 1921, Russia would take advantage of Georgia's weakened position and sweep the little republic back into the empire.

Despite the misgivings of his fellow Georgians, in 1993 Shevardnadze pursued talks toward a comprehensive bilateral Georgian-Russian treaty of friendship. Discussions were interrupted by surges of fighting in Abkhazia, however, and relations were cooled by Shevardnadze's claim that Russia was aiding the secessionist campaign that had begun in August.

In September 1993, the fall of Sukhumi to Abkhazian forces signaled the crumbling of the Georgian army, and the return of Gamsakhurdia threatened to split Georgia into several parts. Shevardnadze, recognizing the necessity of outside military help to maintain his government, agreed to join the CIS on terms dictated by Russia in return for protection of government supply lines by Russian troops. Meanwhile, despite denials by the Yeltsin government, an unknown number of Russians still gave "unofficial" military advice and matériel to the Abkhazian forces, which experts believed would not have posed a major threat to Tbilisi without such assistance. Shevardnadze defended CIS membership at home as an absolute necessity for Georgia's survival as well as a stimulant to increased trade with Russia.

Turkey

Despite a history of episodic Turkish invasions, Shevardnadze courted Turkey as an economic and diplomatic partner. Georgians took advantage of the opening of border traffic with Turkey to begin vigorous commercial activities with their nearest "capitalist" neighbor. In 1992 Georgia became a member of the Black Sea Economic Cooperation Organization, which is based in Turkey. In December 1992, Turkey granted Georgia a credit equivalent to US$50 million to purchase wheat and other goods and to stimulate Turkish private investment in the republic. Georgia also signed several diplomatic agreements with Turkey in the early 1990s, including a Georgian pledge to respect existing common borders, and official Turkish support of Georgian national integrity against the Abkhazian separatist movement. The issue of reinstatement of exiled Meskhetian Turks eased in 1993 when Georgia established official contacts with that minority (see Human Rights).

NATIONAL SECURITY

Military forces have played a critical role in Georgian politics since 1989. In 1991 Georgia's president was overthrown by military force, and the Shevardnadze regime relied heavily on the armed forces to stay in power. Warfare in the autonomous regions of South Ossetia and Abkhazia, as well as armed resistance by Gamsakhurdia supporters in western Georgia, have further emphasized the military's major role in national security.

The Military Establishment

Almost from its inception, the National Guard became directly involved in Georgian politics. By 1992 repeated human rights offenses against Gamsakhurdia supporters brought calls to change this role. At the same time, the political rivalry between Ioseliani and Kitovani, the leaders of the Mkhedrioni (horsemen) and the National Guard, respectively, became one of the key conflicts in the Georgian government hierarchy, and many political parties continued to retain private armies in the guise of armed bodyguards or security teams. Discipline problems in the ranks of both the National Guard and the Mkhedrioni and their ineffectiveness as fighting forces led the Georgian government to plan for a professional army. In April 1992, the State Council adopted a resolution to form a unified armed force of up to 20,000 soldiers.

At the time the government announced its plans for a professional army, however, neither existing military group had sufficient internal discipline to carry out major restructuring. Efforts to disband the National Guard and Mkhedrioni were delayed by continued violence in western Georgia, by an attempted coup in Tbilisi by Gamsakhurdia supporters, and by the political ambitions of Kitovani and Ioseliani. In May 1992, Kitovani was designated minister of defense in an effort to bring the National Guard under central control. Instead, during the following year Kitovani turned his position into a power center rivaling Shevardnadze's. In May 1993, Shevardnadze induced Kitovani and Ioseliani to resign from their powerful positions on the Council for National Security and

Defense, depriving both men of influence over national security policy and enhancing the stature of the head of government.

Shevardnadze complained in early 1993 that a unified army had still not been created. In May the National Guard was abolished as a separate force, and individual distinguished units received guard status. In the second half of 1993, however, outside threats to national security caused Shevardnadze to rely once again on Ioseliani's paramilitary Mkhedrioni, delaying consolidation of a national military force. In September Shevardnadze's control over the military improved when parliament declared a two-month state of emergency that had the effect of weakening the Mkhedrioni.

The Russian Presence

The Soviet Union had maintained a substantial military presence in Georgia because the republic bordered Turkey, a NATO member. The Transcaucasus Military District, which had coordinated Soviet military forces in the three republics of Transcaucasia, was headquartered in Tbilisi. In mid-1993 an estimated 15,000 Russian troops and border guards remained on Georgian territory. Georgia did not press Russian withdrawal as vigorously as did other former republics of the Soviet Union because it did not have enough personnel to patrol its entire border. At the same time, the continued presence of Russian troops energized the Georgian nationalist parties. In the fall of 1993, those groups saw Shevardnadze's call for Russian military assistance, and the significant increase of Russia forces that resulted, as an admission that his national security policy had failed and a sign that the traditional enemy to the north was again threatening.

Draft Policy

The role of Soviet military and internal security forces in the April Tragedy made Georgian connections with those forces a primary target of anticommunist groups. As in other Soviet republics, opposition to the draft became an early focus of opposition activities. Of all the Soviet republics, Georgia had the lowest rate of recruitment in the fall of 1990, approximately 10 percent of eligible citizens. One of the first acts passed by Gamsakhurdia's parliament ended the Soviet military draft on Georgian territory.

In late 1990, Soviet conscription was replaced with the induction of eligible Georgian males into new "special divisions," under the control of the Georgian Ministry of Internal Affairs, for the maintenance of order within the republic. The new body, which became Kitovani's National Guard, was one of the first official non-Soviet military units in what was still the Soviet Union.

Arms Supply

Relatively little of the military industry of the Soviet Union was located in Georgia. One Tbilisi plant assembled military training aircraft that were the basis of a small Georgian air force. Most weapons obtained by the various armed units operating in Georgia after 1990 apparently were purchased illegally from Soviet (and later Russian) officers and soldiers stationed in the Caucasus. In May 1992, leaders of the CIS set quotas for the transfer of Soviet military equipment to republic armed forces. According to this plan, Georgia was to receive 220 tanks, 220 armored vehicles, 300 artillery pieces, 100 military aircraft, and fifty attack helicopters. Kitovani complained in December 1992 that Georgia, unlike the other republics, had not yet received any of its allotment.

Internal Security

The Georgian internal security agency with the closest ties to Moscow was the Georgian branch of the Committee for State Security (Komitet gosudarstvennoi bezopasnosti--KGB). Beginning in 1990, the anticommunist independence movement exerted direct pressure on the Georgian KGB to accept independence. The first confrontation between Moscow and the Gamsakhurdia government came over appointments to top security posts in the republic. In November 1990, the Georgian parliamentary Commission on Security broke the tradition of Moscow-designated KGB chiefs by naming its own appointee. When Gorbachev threatened dire consequences, Gamsakhurdia simply left the chairmanship vacant but named his candidate first deputy chairman and thus acting chairman. At that point, top Georgian KGB officials voiced support for Gamsakhurdia and protested Gorbachev's interference, signaling a service commitment to Tbilisi rather than Moscow.

As late as mid-1991, Moscow continued financing activities of the Georgian KGB and provided part of the budget of the Georgian Ministry of Internal Affairs, which ran domestic intelligence and police agencies. Meanwhile, by 1991 the opposition to Gamsakhurdia was accusing the president of using the Georgian KGB to investigate and harass political enemies.

In May 1992, the Georgian KGB, which in the interim had been renamed the Ministry of Security, was formally replaced by the Information and Intelligence Service. The new agency, formed on the organizational foundation of the old KGB, was headed by Irakli Batiashvili, a thirty-year-old philosophy scholar who had been a National Democratic Party delegate to the National Congress.

Civilian National Security Organization

In November 1992, the parliament passed a law creating the Council for National Security and Defense. This body was accountable to parliament, but, as head of government and commander in chief of the armed forces, Shevardnadze was council chairman. Shevardnadze named Ioseliani and Kitovani deputy chairmen of the council; Tedo Japaridze, top expert on the United States in the Georgian Ministry of Foreign Affairs, became the chairman's aide. The powers of the council included the right to issue binding decisions on military and security matters.

In May 1993, Shevardnadze disbanded the council to deprive Ioseliani and Kitovani of their government power bases. The council was then reconstituted with Shevardnadze's chairmanship assuming greater power.

Crime

In the first postcommunist years, levels of crime and civil unrest in Georgia were quite high because of the proximity of the Armenian-Azerbaijani conflict, refugee movement and terrorism resulting from the Abkhazian conflict within Georgia, the gap between official wages and living standards, and the government's lack of police authority in many areas of the country. Crime statistics were unreliable, however, because the extent of law enforcement and reporting varied during 1993. Reported crimes dropped from 1,982 in May to 1,260 in July. In late 1993, however, numerous automobile thefts and kidnappings occurred on Georgian highways, and citizen insecurity prompted the proliferation of private detective agencies.

The natural gas pipeline to Armenia was a frequent target of terrorist bombs in 1993, and several government figures apparently were the targets of unsuccessful bomb attacks. The Mkhedrioni, who often were involved in criminal activity, usually escaped police control because the minister of internal affairs was a Mkhedrioni member. In September, Shevardnadze took personal control of the ministry to bolster police authority.

Long-Term Security

In late 1993, the primary consideration of Georgian national security continued to be the prevention of territorial gains by separatist national movements--a cause for which Russian military assistance was proving indispensable. Longer-term national security, however, would depend on Shevardnadze's ability to reestablish the structures of a viable, unified state: internal and international commercial activity, undisputed sovereignty over the national territory and its populace, and a shift back to government rule by statute rather than by emergency executive powers. In early 1994, all those preconditions remained in doubt, and Shevardnadze's reluctant resort to Russian military assistance had set a precedent with unknown national security consequences.

*   *   *

For background on Georgian history, the best basic source is Ronald G. Suny's *The Making of the Georgian Nation*. Earlier histories on the Georgian people were written by David Marshall Lang (*A Modern History of Soviet Georgia* and *The Georgians*) and Kalistrat Salia (*History of the Georgian Nation*). Several scholars have followed contemporary Georgian developments on a regular basis; in addition to the present author, they include Elizabeth Fuller, a writer for the *RFE/RL Research Report*; Stephen Jones, whose journal articles cover political and nationalist issues in the Caucasus; and Robert Parsons of the British Broadcasting Corporation. Human rights issues in Georgia are covered extensively in publications of the United States Congress's Commission on Security and Cooperation in Europe. Useful articles from Russian-language sources are translated in the Foreign Broadcast Information Service's *Daily Report: Soviet Union* (more recently titled *Daily Report: Central Eurasia*). Studies of Georgian culture and history appear occasionally in the *Journal for the Study of Caucasia*. (For further information and complete citations, see Bibliography).

CHRONOLOGY OF IMPORTANT EVENTS

Early History

95-55 B.C.
Armenian Empire reaches greatest size and influence under Tigran the Great.

66 B.C.
Romans complete conquest of Caucasus Mountains region, including Georgian kingdom of Kartli-Iberia.

30 B.C.
Romans conquer Armenian Empire.

A.D. 100-300
Romans annex Azerbaijan and name it Albania.

ca. 310
Tiridates III accepts Christianity for the Armenian people.

330
King Marian III of Kartli-Iberia accepts Christianity for the Georgian people.

Fifth-Seventh Centuries

First golden age of Armenian culture.

ca. 600
Four centuries of Arab control of Azerbaijan begin, introducing Islam in seventh century.

645
Arabs capture Tbilisi.

653
Byzantine Empire cedes Armenia to Arabs.

Ninth-Tenth Centuries

806
Arabs install Bagratid family to govern Armenia.

813
Armenian prince Ashot I begins 1,000 years of rule in Georgia by Bagratid Dynasty.

862-977
Second golden age of Armenian culture, under Ashot I and Ashot III.

Eleventh-Fourteenth Centuries

Byzantine Greeks invade Armenia from west, Seljuk Turks from east; Turkish groups wrest political control of Azerbaijan from Arabs, introducing Turkish language and culture.

1099-1125
David IV the Builder establishes expanded Georgian Empire and begins golden age of Georgia.

1000-late 1200s
Golden age of Azerbaijani literature and architecture.

1100s-1300s
Cilician Armenian and Georgian armies aid European armies in Crusades to limit Muslim control of Holy Land.

1200-1400
Mongols twice invade Azerbaijan, establishing temporary dynasties.

1375
Cilician Armenia conquered by Mamluk Turks.

1386
Timur (Tamerlane)
sacks Tbilisi, ending Georgian Empire

Fifteenth Century

Most of modern Armenia, Azerbaijan, and Georgia become part of Ottoman Empire.

Sixteenth Century

1501
Azerbaijani Safavid Dynasty begins rule by Persian Empire.

1553
Ottoman Turks and Persians divide Georgia between them.

Eighteenth Century

ca. 1700
Russia begins moving into northern Azerbaijan as Persian Empire
weakens.

1762
Herekle II reunites eastern Georgian regions in kingdom of Kartli-Kakhetia.

Nineteenth Century

1801
After Herekle II's appeal for aid, Russian Empire abolishes Bagratid Dynasty and
begins annexation of Georgia.

1811
Georgian Orthodox Church loses autocephalous status in Russification process.

1813
Treaty of Gulistan officially divides Azerbaijan into Russian (northern) and Persian (southern) spheres.

1828
Treaty of Turkmanchay awards Nakhichevan and area around Erevan to Russia, strengthening Russian control of Transcaucasus and beginning period of modernization and security.

1872
Oil industry established around Baku, beginning rapid expansion.

1878
"Armenian question" emerges at Congress of Berlin; disposition of Armenia becomes ongoing European issue.

1891
First Armenian revolutionary party formed.

1895
Massacre of 300,000 Armenian subjects by Ottoman Turks.

Twentieth Century

ca. 1900
Radical political organizations begin to form in Azerbaijan.

1908
Young Turks take over government of Ottoman Empire with reform agenda, supported by Armenian population.

1915

Young Turks massacre 600,000 to 2 million Armenians; most survivors leave eastern Anatolia.

1917

Armenia, Azerbaijan, and Georgia form independent Transcaucasian federation. Tsar Nicholas II abdicates Russian throne; Bolsheviks take power in Russia.

1918

Independent Armenian, Azerbaijani, and Georgian states emerge from defeat of Ottoman Empire in World War I.

1920

Red Army invades Azerbaijan and forces Armenia to accept communist-dominated government.

1921

Red Army invades Georgia and drives out Zhordania government.

1922

Transcaucasian Soviet Federated Socialist Republic combines Armenia, Azerbaijan, and Georgia as single republic within Soviet Union.

1936

Armenia, Azerbaijan, and Georgia become separate republics within Soviet Union.

1936-37

Purges under political commissar Lavrenti Beria reach their peak in Armenia, Azerbaijan, and Georgia.

1943

Autonomy restored to Georgian Orthodox Church.

1946

Western powers force Soviet Union to abandon Autonomous Government of Azerbaijan, formed in 1945 after Soviet occupation of northern Iran.

1959

Nikita S. Khrushchev purges Azerbaijani Communist Party.

1969

Heydar Aliyev named head of Azerbaijani Communist Party.

ca. 1970

Zviad Gamsakhurdia begins organizing dissident Georgian nationalists.

1972

Eduard Shevardnadze named first secretary of Georgian Communist Party.

1974

Moscow installs regime of Karen Demirchian in Armenia to end party corruption; regime later removed for corruption.

1978

Mass demonstrations prevent Moscow from making Russian an official language of Georgia.

1982

Aliyev of Azerbaijan named full member of Politburo of Communist Party of the Soviet Union.

1985

Shevardnadze named minister of foreign affairs of Soviet Union and leaves post as first secretary of Georgian Communist Party.

Late 1980s

Mikhail S. Gorbachev initiates policies of *glasnost* and *perestroika* throughout Soviet Union.

1988
Armenian nationalist movement revived by Karabakh and corruption concerns.

February
Nagorno-Karabakh government votes to unify that autonomous region of Azerbaijan with Armenia.

December
Disastrous earthquake in northern Armenia heavily damages Leninakan (now Gyumri).

1989
April
Soviet troops kill Georgian civilian demonstrators in Tbilisi, radicalizing Georgian public opinion.

Spring
Mass demonstrations in Armenia achieve release of Karabakh Committee arrested by Soviets to quell nationalist movement.

September
Azerbaijan begins blockade of Armenian fuel and supply lines over Karabakh issue.

Fall
Azerbaijani opposition parties lead mass protests against Soviet rule; national sovereignty officially proclaimed.

November
Nagorno-Karabakh National Council declares unification of Nagorno-Karabakh with Armenia.

1990
January
Moscow sends troops to Azerbaijan, nominally to stem violence against Armenians over Karabakh.

Spring
Levon Ter-Petrosian of Armenian Pannational Movement chosen chairman of Armenian Supreme Soviet.

October
In first multiparty election held in Georgia, Gamsakhurdia's oppositionist party crushes communists; Gamsakhurdia named president.

1991
January
Georgian forces invade South Ossetia in response to independence movement there; fighting continues all year; Soviet troops invade Azerbaijan, ostensibly to halt anti-Armenian pogroms.

April
After referendum approval, Georgian parliament declares Georgia independent of Soviet Union.

May
Gamsakhurdia becomes first president of Georgia, elected directly in multiparty election.

August
Attempted coup against Gorbachev in Moscow fails.

September
Armenian voters approve national independence.

October
Azerbaijani referendum declares Azerbaijan independent of Soviet Union; Ter-Petrosian elected president of Armenia.

December
Armenians in Nagorno-Karabakh declare independent state as fighting there continues; Soviet Union officially dissolved.

1992
January
Gamsakhurdia driven from Georgia into exile by opposition forces.

March
Shevardnadze returns to Tbilisi and forms new government.

Spring
Armenian forces occupy Lachin corridor linking Nagorno-Karabakh to Armenia.

June
Abulfaz Elchibey elected president of Azerbaijan and forms first postcommunist government there.

July
Cease-fire mediated by Russia's President Yeltsin in South Ossetia.

October
Parliamentary election held in Georgia; Shevardnazde receives overwhelming support.

Fall
Fighting begins between Abkhazian independence forces and Georgian forces; large-scale refugee displacement continues through next two years.

June
Military coup deposes Elchibey in Azerbaijan; Aliyev returns to power.

Fall
Multilateral negotiations seek settlement of Karabakh conflict, without result; fighting, blockade, and international negotiation continue into 1994.

October
Shevardnadze responds to deterioration of Georgian military position by having Georgia join Commonwealth of Independent States, thus gaining Russian military support; Aliyev elected president of Azerbaijan.

GLOSSARY -- GEORGIA (CAUCASUS)

Commonwealth of Independent States (CIS)
Official designation of the former republics of the Soviet Union that remained loosely federated in economic and security matters of common concern after the Soviet Union disbanded as a unified state in 1991. Members in early 1994 were Armenia, Azerbaijan, Belarus, Georgia, Kazakhstan, Kyrgyzstan, Moldova, Russia, Tajikistan, Turkmenistan, Ukraine, and Uzbekistan.

Communist Party of the Soviet Union (CPSU)
The official name of the communist party in the Soviet Union after 1952. Originally the Bolshevik (majority) faction of a prerevolutionary Russian party, the party was named the Russian Communist Party (Bolshevik) from 1918 until it was renamed in 1952.

Conference on Security and Cooperation in Europe (CSCE)
Originating in Helsinki in 1975, a grouping of all European nations (the only exception, Albania, joined in 1991) that has sponsored joint sessions and consultations on political issues vital to European security.

Conventional Forces in Europe Treaty (CFE Treaty)
An agreement signed in 1990 by the members of the Warsaw Treaty Organization (Warsaw Pact--*q.v.*) and the North Atlantic Treaty Organization (NATO--*q.v.*) to establish parity in conventional weapons between the two organizations from the Atlantic to the Urals. Included a strict system of inspections and information exchange.

coupon
Generic term for bank-issued national currency certificates of Georgia, introduced in early 1993. After introduction, value declined rapidly; in October 1993, the exchange rate was approximately 42,000 coupons per US$1.

dram
National currency of Armenia, officially established for use concurrent with the Russian ruble in November 1993, to become single official currency in early 1994. In November 1993, the exchange rate was approximately 14 drams per US$1. A second national unit, the luma (100 to the dram), was introduced in February 1994.

glasnost
Russian term, literally meaning "openness." Applied in the Soviet Union beginning in the mid-1980s to official permission for public discussion of issues and public access to information. Identified with the tenure of Mikhail S. Gorbachev as leader of the Soviet Union.

gross domestic product (GDP)
The total value of goods and services produced exclusively within a nation's domestic economy, in contrast to the gross national product (*q.v.*). Normally computed over one-year periods.

gross national product (GNP)
The total value of goods and services produced within a country's borders and the income received from abroad by residents, minus payments remitted abroad by nonresidents. Normally computed over one-year periods.

International Monetary Fund (IMF)
Established in 1945, a specialized agency affiliated with the United Nations and responsible for stabilizing international exchange rates and payments. Its main business is providing loans to its members when they experience balance of payments difficulties.

Kurds
A mainly Muslim people speaking an Indo-European language similar to Persian. Kurds constitute significant minorities in Iran, Iraq, and Turkey, with smaller groups in Armenia and Syria. Despite international proposals in response to minority persecution, never united in a single state.

manat
National currency of Azerbaijan. Introduced in mid-1992 for use concurrent with the ruble; became sole official currency in January 1994. In October 1993, the exchange rate was 120 manats per US$1.

millet
In the Ottoman Empire, the policy for governance of non-Muslim minorities. The system created autonomous communities ruled by religious leaders responsible to the central government.

net material product (NMP)
In countries having centrally planned economies, the official measure of the value
of goods and services produced within the country. Roughly equivalent to the
gross national product (*q.v.*), NMP is based on constant prices and does not
account for depreciation.

North Atlantic Treaty Organization (NATO)
During the postwar period until the dissolution of the Soviet Union in 1991, the
primary collective defense agreement of the Western powers against the
military presence of the Warsaw Pact (*q.v.*) nations in Europe. Founded 1949.
Its military and administrative structure remained intact after 1991, but early
in 1994 the Partnership for Peace proposed phased membership to all East
European nations and many former republics of the Soviet Union.

perestroika
Russian term meaning "restructuring." Applied in the late 1980s to an official
Soviet program of revitalization of the Communist Party of the Soviet Union
(CPSU--*q.v.*), the economy, and the society by adjusting economic, social,
and political mechanisms in the central planning system. Identified with the
tenure of Mikhail S. Gorbachev as leader of the Soviet Union.

Shia
The smaller of the great two divisions of Islam, supporting the claims of Ali to
leadership of the Muslim community, in opposition to the Sunni (*q.v.*) view of
succession to Muslim leadership--the issue causing the central schism within
Islam.

Sunni
The larger of the two fundamental divisions of Islam, opposed to the Shia (*q.v.*)
on the issue of succession of Muslim leadership.

value-added tax (VAT)
A tax applied to the additional value created at a given stage of production and
calculated as a percentage of the difference between the product value at that
stage and the cost of all materials and services purchased or introduced as
inputs.

Volunteer Society for Assistance to the Army, Air Force, and Navy (DOSAAF)
In the Soviet national defense system, the agency responsible for paramilitary training of youth and reserve components.

Warsaw Pact
Informal name for the Warsaw Treaty Organization, a mutual defense organization founded in 1955. Included the Soviet Union, Albania (which withdrew in 1968), Bulgaria, Czechoslovakia, the German Democratic Republic (East Germany), Hungary, Poland, and Romania. The Warsaw Pact enabled the Soviet Union to station troops in the countries of Eastern Europe to oppose the forces of the North Atlantic Treaty Organization (NATO--*q.v.*). The pact was the basis for the invasions of Hungary (1956) and Czechoslovakia (1968). Disbanded in July 1991.

World Bank
Informal name for a group of four affiliated international institutions: the International Bank for Reconstruction and Development (IBRD); the International Development Association (IDA); the International Finance Corporation (IFC); and the Multilateral Investment Guarantee Agency (MIGA). The four institutions are owned by the governments of the countries that subscribe their capital for credit and investment in developing countries; each institution has a specialized agenda for aiding economic growth in target countries. To participate in the World Bank group, member states must first belong to the International Monetary Fund (IMF--*q.v.*).

BIBLIOGRAPHY -- GEORGIA (CAUCASUS)

Afandiiev, Rasim. *Folk Art of Azerbaijan*. Baku: Ishyg, 1984.

Aftandilian, Gregory. *Armenia, Vision of a Republic: The Independence Lobby in America*. Boston: Charles River Books, 1981.

Akiner, Shirin. *Islamic Peoples of the Soviet Union*. (2d ed.) London: Routledge and Kegan Paul, 1986.

Alstadt, Audrey L. "The Azerbaijani Bourgeoisie and the Cultural- Enlightenment Movement in Baku: First Steps Toward Nationalism." In Ronald G. Suny (ed.), *Transcaucasia: Nationalism and Social Change: Essays in the History of Armenia, Azerbaijan, and Georgia*. Ann Arbor: Michigan Slavic Publications, 1983.

Alstadt, Audrey L. *The Azerbaijani Turks: Power and Identity under Russian Rule*. Stanford, California: Hoover Institution Press, 1992.

Apostolou, Andrew, Amberin Zaman, and Naritza Matossian. "Central Asia: New Players in an Old Game," *Middle East*, No. 213, July 1992, 5-10.

Arberry, A.J. (ed.). *Religion in the Middle East*, 1. Cambridge: Cambridge University Press, 1969.

Arlen, Michael. *Passage to Ararat*. New York: Farrar, Straus and Giroux, 1975.

Arpee, Leon. *The Armenian Awakening: A History of the Armenian Church, 1820-1860*. Chicago: University of Chicago Press, 1909.

Aslan, Kevork. *Armenia and the Armenians from the Earliest Times until the Great War (1914)*. (Trans., Pierre Crabites.) New York: Macmillan, 1920.

Aspaturian, Vernon V. *The Union Republics in Soviet Diplomacy: A Study of Soviet Federalism in the Service of Soviet Foreign Policy*. Geneva: E. Droz, 1960.

Atamian, Sarkis. *The Armenian Community: The Historical Development of a Social and Ideological Conflict*. New York: Philosophical Library, 1955.

Bardakjian, Kevork B. *The Mekhitarist Contributions to Armenian Culture and Scholarship*. Cambridge: Harvard College Library, 1976.

Bauer-Manndorff, Elisabeth. *Armenia: Past and Present*. Lucerne: Reich Verlag, 1981.

Benet, Sula. *Abkhazians: The Long-Living People of the Caucasus*. New York: Holt, Rinehart and Winston, 1974.

Bennigsen, Alexandre, and Marie Broxup. *The Islamic Threat to the Soviet State*. London: Croom Helm, 1983.

Bennigsen, Alexandre, Marie Broxup, and S. Enders Wimbush. *Muslims of the Soviet Empire: A Guide*. Bloomington: Indiana University Press, 1986.

Blackwell, Alice Stone. *Armenian Poetry*. Boston: Atlantic, 1917.

Bournoutian, George A. *Eastern Armenia in the Last Decades of Persian Rule, 1807-1828: A Political and Socioeconomic Study of the Khanate of Erevan on the Eve of the Russian Conquest.* Malibu, California: Undena, 1982.

Braude, Benjamin, and Bernard Lewis. *Christians and Jews in the Ottoman Empire: The Functioning of a Plural Society.* (2 vols.) New York: Holmes and Meier, 1982.

Bremmer, Ian, and Ray Taras (eds.). *Nations and Politics in the Soviet Successor States.* Cambridge: Cambridge University Press, 1993.

Bryce, James. *Transcaucasia and Ararat.* London: Macmillan, 1896. Reprint. New York: Arno Press, 1970.

Buxton, Noel, and Harold Buxton. *Travels and Politics in Armenia.* London: Murray, 1914.

Chalabian, Antranig. *General Andranik and the Armenian Revolutionary Movement.* Southfield, Michigan: Chalabian, 1988.

Dale, Catherine. "Turmoil in Abkhazia: Russian Responses," *RFE/RL Research Report* [Munich], 2, No. 34, August 27, 1993, 48-57.

"Dangerous Liaisons in Transcaucasia," *Intelligence Digest,* February 18, 1994, 3-4.

Davis, Christopher M. "Health Care Crisis: The Former Soviet Union," *RFE/RL Research Report* [Munich], 2, No. 40, October 8, 1993, 35-43.

Davison, Roderic. "The Armenian Crisis, 1912-1914," *American Historical Review,* 53, No. 3, April 1948, 481-505.

Davison, Roderic. *Reform in the Ottoman Empire, 1856-1876.* Princeton: Princeton University Press, 1963.

de Morgan, Jacques. *History of the Armenian People.* (Trans., Ernest F. Barry.) Boston: Hairenik Press, n.d.

Dekmejian, R.H. "Soviet-Turkish Relations and Politics in the Armenian SSR," *Soviet Studies,* 19, No. 4, April 1968, 510-25.

des Pres, Terrence. "On Governing Narratives: The Turkish-Armenian Case," *Yale Review,* 75, Summer 1986, 517-31.

Dragadze, Tamara. "Azerbaijanis." Pages 163-79 in Graham Smith (ed.), *The Nationalities Question in the Soviet Union.* London: Longman, 1990.

Dragadze, Tamara. "Conflict in the Transcaucasus and the Value of Inventory Control," *Jane's Intelligence Review* [London], 6, No. 2, February 1994, 71-3.

Etmekjian, James. *The French Influence on the Western Armenian Renaissance, 1843-1915.* New York: Twayne, 1964.

Fawcett, Louise l'Estrange. *Iran and the Cold War: The Azerbaijan Crisis of 1946.* Cambridge: Cambridge University Press, 1992.

Frelick, Bill. *Faultlines of Nationality Conflict: Refugees and Displaced Persons from Armenia and Azerbaijan*. Washington: U.S. Committee for Refugees, 1994.

Fuller, Elizabeth. "Armenia's Constitutional Debate," *RFE/RL Research Report* [Munich], 3, No. 22, May 27, 1994, 6-9.

Fuller, Elizabeth. "Azerbaijan's June Revolution," *RFE/RL Research Report* [Munich], 2, No. 32, August 13, 1993, 24- 29.

Fuller, Elizabeth. "Azerbaijan's Relations with Russia and the CIS," *RFE/RL Research Report* [Munich], 1, No. 43, October 30, 1992, 52-55.

Fuller, Elizabeth. "Eduard Shevardnadze's Via Dolorosa," *RFE/RL Research Report* [Munich], 2, No. 43, October 29, 1993, 17-23.

Fuller, Elizabeth. "Ethnic Strife Threatens Democratization," *RFE/RL Research Report* [Munich], 2, No. 1, January 1, 1993, 17- 24.

Fuller, Elizabeth. "Georgia since Independence: Plus Ca Change . . .," *Current History*, 92, October 1993, 342-46.

Fuller, Elizabeth. "Georgia, Abkhazia, and Checheno-Ingushetia," *RFE/RL Research Report* [Munich], 1, No. 5, February 5, 1992, 3- 7.

Fuller, Elizabeth. "Paramilitary Forces Dominate Fighting in Transcaucasus," *RFE/RL Research Report* [Munich], 2, No. 25, June 18, 1993, 74-82.

Fuller, Elizabeth. "Russia, Turkey, Iran, and the Karabakh Mediation Process," *RFE/RL Research Report* [Munich], 3, No. 8, February 25, 1994, 31-36.

Fuller, Elizabeth. "The Georgian Parliamentary Elections," *RFE/RL Research Report* [Munich], 1, No. 47, November 27, 1992, 1-4.

Fuller, Elizabeth. "The Karabakh Mediation Process: Grachev Versus the CSCE?" *RFE/RL Research Report* [Munich], 3, No. 23, June 10, 1994, 13-17.

Gidney, James B. *A Mandate for Armenia*. Kent, Ohio: Kent State University Press, 1967.

Goble, Paul A. "Coping with the Nagorno-Karabakh Crisis," *Fletcher Forum of World Affairs*, 16, Summer 1992, 19- 26.

Golden, Peter. "The Turkic Peoples and Transcaucasia." Pages 45-68 in Ronald G. Suny (ed.), *Transcaucasia: Nationalism and Social Change: Essays in the History of Armenia, Azerbaijan, and Georgia*. Ann Arbor: Michigan Slavic Publications, 1983.

Goshgarian, Geoffrey. "Eghishe Charents and the `Modernization' of Soviet Armenian Literature," *Armenian Review*, 36, No. 1, 1983, 76-88.

Hartunian, Abraham H. *Neither to Laugh Nor to Weep: A Memoir of the Armenian Genocide*. (Trans., Vartan Hartunian.) Boston: Beacon Press, 1968.

Henry, J.D. *Baku: An Eventful History*. London: Constable, 1905. Reprint. New York: Arno Press, 1977.

Holisky, Dee Ann. "The Rules of the *Supra* or How to Drink in Georgian," *Journal for the Study of Caucasia*, 1, No. 1, 1-20.

Hostler, Charles W. *Turkism and the Soviets*. London: Allen and Unwin, 1957.

Hovannisian, Richard G. "Russian Armenia: A Century of Tsarist Rule," *Jahrbücher für Geschichte Osteuropas* [Munich], 19, No. 1, 1971, 31-48.

Hovannisian, Richard G. (ed.). *The Armenian Genocide in Perspective*. New Brunswick, New Jersey: Transaction Books, 1986.

Hovannisian, Richard G. (ed.). *The Armenian Image in History and Literature*. Malibu, California: Undena Press, 1981.

Hovannisian, Richard G. *The Armenian Holocaust: A Bibliography Relating to the Deportations, Massacres, and Dispersion of the Armenian People, 1915-1923*. Cambridge, Massachusetts: Armenian Heritage Press, 1978.

Hovannisian, Richard G. *The Armenian Republic*, 1 and 2. Berkeley:

Hovannisian, Richard G., Stanford J. Shaw, and Ezel Kural. "Forum: The Armenian Question," *International Journal of Middle East Studies*, 11, No. 3, August 1979, 379-400.

Huddle, Frank, Jr. "Azerbaidzhan and the Azerbaidzhanis," Pages 189-210 in Zev Katz (ed.), *Handbook of Major Soviet Nationalities*. New York: Free Press, 1975.

Ibrahimov, Mirza (ed.). *Azerbaijanian Poetry: Classic, Modern, Traditional*. Moscow: Progress, 1969.

International Monetary Fund. *Armenia*. (IMF Economic Reviews, 1/1993.) Washington: 1993.

International Monetary Fund. *Economic Review: Azerbaijan*. Washington: 1992.

International Monetary Fund. *Economic Review: Georgia*. Washington: 1992.

International Petroleum Encyclopedia, 1992. (Ed., Jim West.) Tulsa: PennWell, 1992.

Jane's World Railways, 1993-94. (Ed., James Abbott.) Alexandria, Virginia: Jane's Information Group, 1993.

Jones, Stephen. "A Failed Democratic Transition." Pages 288-310 in Ian Bremmer and Ray Taras (eds.), *Nations and Politics in the Soviet Successor States*. Cambridge: Cambridge University Press, 1993.

Jones, Stephen. "Religion and Nationalism in Soviet Georgia and Armenia." Pages 171-95 in Pedro Ramet (ed.), *Religion and Nationalism in Soviet and East European Politics*. Durham: Duke University Press, 1989.

Jones, Stephen. "The Caucasian Mountain Railway Project: A Victory for *Glasnost?*" *Central Asian Survey*, 8, No. 2, 1989, 47-59.

Katz, Zev (ed.). *Handbook of Major Soviet Nationalities*. New York: Free Press, 1975.

Katz, Zev. "The Armenian Church and the WCC: A Personal View," *Ecumenical Review*, 40, July-October 1988, 411-17.

Kazemzadeh, Firuz. *The Struggle for Transcaucasia, 1917- 1921*. New York: Philosophical Library, 1951.

Lang, David Marshall, and Christopher J. Walker. *The Armenians*. (Minority Rights Group Report No. 32.) London: Minority Rights Group, 1976.

Lang, David Marshall. *A Modern History of Soviet Georgia*. New York: Grove Press, 1962.

Lang, David Marshall. *Armenia: Cradle of Civilization*. London: Allen and Unwin, 1980.

Lang, David Marshall. *The Armenians: A People in Exile*. London: Allen and Unwin, 1981.

Lang, David Marshall. *The Georgians*. London: Thames and Hudson, 1966.

Libaridian, Gerard J. *The Karabagh File: Documents and Facts on the Question of Mountainous Karabagh, 1918-1988*. Cambridge, Massachusetts: Zoryan Institute, 1988.

Lloyd's Ports of the World, 1988. (Ed., Paul J. Cuny.) Colchester, United Kingdom: Lloyd's of London, 1988.

Lynch, H.F.B. *Armenia: Travels and Studies*. (2 vols.) Beirut: Khayats, 1965.

Maggs, William Ward. "Armenia and Azerbaijan: Looking Toward the Middle East," *Current History*, 92, January 1993, 6- 11.

Mars, Gerald, and Yochanan Altman. "The Cultural Bases of Soviet Georgia's Second Economy," *Soviet Studies*, 35, No. 4, October 1983, 546-60.

Matossian, Mary Kilbourne. *The Impact of Soviet Policies in Armenia*. Leiden, Netherlands: E.J. Brill, 1962.

Mirak, Robert. "Armenians." Pages 136-49 in Steven Thernstrom (ed.), *Harvard Encyclopedia of American Ethnic Groups*. Cambridge: Harvard University Press, 1980.

Nalbandian, Louise. *The Armenian Revolutionary Movement: The Development of Armenian Political Parties Through the Nineteenth Century*. Berkeley: University of California Press, 1963.

National Democratic Movement in Armenia. *Armenia at the Crossroads: Democracy and Nationhood in the Post-Soviet Era*. Watertown, Massachusetts: Blue Crane, 1991.

Nichol, James. "Georgia in Transition: Context and Implications for U.S. Interests," *Congressional Research Service Report*. (No. 93794.) Washington: Library of Congress, Congressional Research Service, August 24, 1993.

Nissman, David. "The National Reawakening of Azerbaijan," *World and I*, 7, February 1992, 80-85.

Noble, John S., and John King. *USSR: A Travel Survival Kit.* Berkeley, California: Lonely Planet, 1993.

Parsons, Robert. "Georgians." Pages 180-96 in Graham Smith (ed.), *The Nationalities Question in the Soviet Union.* New York: Longman, 1990.

Pipes, Richard. *The Formation of the Soviet Union: Communism and Nationalism, 1917-1923.* New York: Atheneum, 1968.

Pipes, Richard. *The Formation of the Soviet Union: Communism and Nationalism, 1917-1923.* Cambridge: Harvard University Press, 1964.

Rakowska-Harmstone, Teresa. "The Dialectics of Nationalism in the USSR," *Problems of Communism,* 13, No. 3, 1974, 1-22.

Ramet, Pedro (ed.). *Religion and Nationalism in Soviet and East European Politics.* Durham: Duke University Press, 1989.

Richards, Susan. *Epics of Everyday Life: Encounters in a Changing Russia.* New York: Viking Press, 1990.

Rosen, Roger. *The Georgian Republic.* Lincolnwood, Illinois: Passport Books, 1992.

Rustaveli, Shota. *The Knight in the Panther's Skin.* (Trans., Venera Urushadze.) Tbilisi: Sabchota Sakartvelo, 1986.

Salia, Kalistrat. *History of the Georgian Nation.* (Trans., Katharine Vivian.) (2d ed.) Paris: Académie Française, 1983.

Sarkissian, Karekin. "The Armenian Church." Pages 482-520 in A.J. Arberry (ed.), *Religion in the Middle East,* 1. Cambridge: Cambridge University Press, 1969.

Sarkisyanz, Manuel. *A Modern History of Transcaucasian Armenia: Social, Cultural, and Political.* Leiden, Netherlands: E.J. Brill, 1975.

Saroyan, Mark. "The `Karabakh Syndrome' and Azerbaijani Politics," *Problems of Communism,* 39, September-October 1990, 14-29.

Sebag-Montefiore, Simon. "Eduard Shevardnadze," *New York Times Magazine,* December 26, 1993, 16-19.

Shaw, Stanford Jay, and Ezel Kural. *History of the Ottoman Empire and Modern Turkey.* (2 vols.) Cambridge: Cambridge University Press, 1977.

Sheehy, Ann, and Elizabeth Fuller. "Armenia and Armenians in the USSR: Nationality and Language Aspects of the Census of 1979," *Radio Liberty Research Bulletin* [Munich], 24, No. 22, June 13, 1980.

Shevardnadze, Eduard. *The Future Belongs to Freedom.* New York: Free Press, 1991.

Slider, Darrell. "Crisis and Response in Soviet Nationality Policy: The Case of Abkhazia," *Central Asian Survey,* 4, No. 4, 1985, 51-68.

Slider, Darrell. "Party-Sponsored Public Opinion Research in the Soviet Union, " *Journal of Politics*, 47, No. 1, February 1985, 209-27.

Slider, Darrell. "The Politics of Georgia's Independence," *Problems of Communism*, 40, No. 6, November-December 1991, 63-79.

Smith, Anthony D. *The Ethnic Origins of Nations*. Oxford: Blackwell, 1986.

Smith, Graham (ed.). *The Nationalities Question in the Soviet Union*. New York: Longman, 1990.

Sobhani, Sohrab C. "Azerbaijan: A People in Search of Independence," *Global Affairs*, 6, Winter 1991, 40-53.

Sturino, John. "Nagorno-Karabakh: The Roots of Conflict," *Surviving Together*, 11, Spring 1993, 10-14.

Suny, Ronald G. "Incomplete Revolution: National Movements and the Collapse of the Soviet Empire," *New Left Review*, No. 189, September-October 1991, 111-26.

Suny, Ronald G. "Nationalism and Democracy in Gorbachev's Soviet Union: The Case of Karabagh," *Michigan Quarterly Review*, 28, No. 4, 481-506.

Suny, Ronald G. "The Revenge of the Past: Socialism and Ethnic Conflict in Transcaucasia," *New Left Review*, No. 184, November-December 1990, 5-34.

Suny, Ronald G. (ed.). *Transcaucasia: Nationalism and Social Change: Essays in the History of Armenia, Azerbaijan, and Georgia*. Ann Arbor: Michigan Slavic Publications, 1983.

Suny, Ronald G. *Armenia in the Twentieth Century*. Chico, California: Scholars Press, 1983.

Suny, Ronald G. *Looking Toward Ararat: Armenia in Modern History*. Bloomington: Indiana University Press, 1993.

Suny, Ronald G. *The Baku Commune, 1917-1918: Class and Nationality in the Russian Revolution*. Princeton: Princeton University Press, 1972.

Suny, Ronald G. *The Making of the Georgian Nation*. Bloomington: Indiana University Press, 1988.

Sutton, Peter. "A View of Soviet Armenia," *Contemporary Review*, No. 246, May 4, 1985, 251-54.

Sweeney, Padraic (ed.). *The Transportation Handbook for Russia and the Former Soviet Union*. Arlington, Virginia: ASET/ICI, 1993.

Swietochowski, Tadeusz. "Azerbaijan: Between Ethnic Conflict and Irredentism," *Armenian Review*, 43, Summer-Autumn 1990, 35-49.

Swietochowski, Tadeusz. "National Consciousness and Political Orientations in Azerbaijan, 1905-1920." Pages 209-38 in Ronald G. Suny (ed.), *Transcaucasia: Nationalism and Social Change: Essays in the History of*

Armenia, Azerbaijan, and Georgia. Ann Arbor: Michigan Slavic Publications, 1983.

Swietochowski, Tadeusz. *Russian Azerbaijan, 1905-1920: The Shaping of National Identity in a Muslim Community.* Cambridge: Cambridge University Press, 1985.

Ter Minassian, Anahide. *La République d'Arménie.* Paris: Éditions Complexes, 1989.

Ter Minassian, Anahide. *Nationalism and Socialism in the Armenian Revolutionary Movement.* Cambridge, Massachusetts: Zoryan Institute, 1984.

The Europa World Year Book, 1993, 1. London: Europa, 1993.

The Europa World Year Book, 1994, 1. London: Europa, 1994.

The Military Balance, 1994-1995. London: Brassey's for International Institute for Strategic Studies, 1994.

The Statesman's Year-Book, 1993-1994. (Ed., Brian Hunter.) New York: St. Martin's Press, 1994.

The Statesman's Year-Book, 1994-1995. (Ed., Brian Hunter.) New York: St. Martin's Press, 1994.

Thernstrom, Steven (ed.). *Harvard Encyclopedia of American Ethnic Groups.* Cambridge: Harvard University Press, 1980.

Thorossian, H. *Histoire de la littérature arménienne: Des origines jusqu'à nos jours.* Paris: n.p., 1951.

Timofeyev, A. *Motorist's Guide to the Soviet Union.* Moscow: Progress, 1980.

Toynbee, Arnold J. *Armenian Atrocities: The Murder of a Nation.* London: Hodder and Stoughton, 1915.

Tremlett, P.I. (ed.). *Thomas Cook Overseas Timetable.* Thorpe Wood, United Kingdom: Thomas Cook, 1994.

Union," *Journal of Politics,* 47, No. 1, February 1985, 209-27.

United States. Agency for International Development. Center for International Health Information. *Armenia: USAID Health Profile.* Arlington, Virginia: 1992.

United States. Central Intelligence Agency. *Handbook of International Economic Statistics, 1993.* Washington: GPO, 1993.

United States. Central Intelligence Agency. *The World Factbook, 1994.* Washington: 1994.

United States. Central Intelligence Agency. *USSR Energy Atlas.* Washington: 1985.

United States. Congress. 103d, 1st Session. Commission on Security and Cooperation in Europe. *Human Rights and Democratization in the Newly Independent States of the Former Soviet Union.* Washington: GPO, 1993.

United States. Department of Commerce. Business Information Service for the Newly Independent States. *Economic Profile of Armenia*. Washington: 1993.

United States. Department of Commerce. Business Information Service for the Newly Independent States. *Georgia*. Washington: 1993.

United States. Department of State. "Armenia," *U.S. Department of State Dispatch*, 5, May 2, 1994, 255-57.

University of California Press, 1971 and 1982.

Walker, Christopher J. "Armenia: A Nation in Asia," *Asian Affairs*, 19, February 1988, 20-35.

Walker, Christopher J. (ed.). *Armenia and Karabagh: The Struggle for Unity*. London: Minority Rights Group, 1991.

Walker, Christopher J. *Armenia: The Survival of a Nation*. New York: St. Martin's Press, 1980.

Weekes, Richard W. (ed.). *The Muslim Peoples: A World Ethnographic Survey*. (2d ed.) Westport, Connecticut: Greenwood Press, 1984.

Woff, Richard. "The Armed Forces of Armenia," *Jane's Intelligence Review* [London], 6, September 1994, 387-91.

World Bank. *Azerbaijan: From Crisis to Sustained Growth*. (A World Bank Country Study.) Washington: 1993.

World Bank. *Food and Agricultural Policy Reforms in the Former USSR*. (Studies of Economies in Transformation, Paper No. 1.) Washington: 1992.

World Bank. *Georgia: A Blueprint for Reforms*. (A World Bank Country Study.) Washington: 1993.

World Bank. *Statistical Handbook: States of the Former USSR*. (Studies of Economies in Transformation, Paper No. 3.) Washington: 1992.

Yefendizade, R.M. *Architecture of the Soviet Azerbaijan*. Moscow: Stroiizdat, 1986.

Various issues of the following periodicals were also used in the preparation of this chapter: Armenian Assembly of America, *Monthly Digest of News from Armenia*; and Foreign Broadcast Information Service, *Daily Report: Central Eurasia*, *FBIS Report: Central Eurasia*, and *JPRS Report: Environmental Issues*.

INDEX

#

1990 parliamentary election, 121
1992 election, 124, 125, 126, 134

A

Abkhaz separatist forces, 19
Abkhaz(ian) conflict, 7, 8, 19, 51, 54, 132, 142
Abkhazia, 3, 6, 7, 11, 13, 18-20, 30, 36, 51, 53, 54, 58, 59, 75, 82-87, 89, 104, 108, 113, 122, 126, 131, 136-139, 158, 159, 162
Abkhazian Autonomous Republic, 1, 30, 36, 58, 59, 65, 75, 83, 88, 112, 115
Adzharian Autonomous Republic, 1
Afghanistan, 11
agriculture, 17, 33, 44, 75, 109, 127
AIDS (epidemic), 98, 99
Albania, 144, 153, 156
Aliyev, 44-48, 57, 58, 59, 149, 152
Aliyev, Heydar, 39, 41, 44, 58, 149
annexation of Georgia, 2, 146
anti-government rallies, 15
Arabs, 2, 32, 68, 144, 145
architectural style, 93
Armenia, 7, 9, 13, 18, 21, 29-33, 35-50, 54-57, 63, 64, 66, 68, 73, 84, 98, 107, 108, 113, 135-137, 142, 145-154, 157-165
Armenian Apostolic Church, 92
Armenian economy, 37
Armenian Pannational Movement (APM), 41, 150
Armenian-Azerbaijani conflict, 137, 142
assassination attempt, 6
autonomous republic, 83, 88, 89
Azerbaijan, 7, 8, 13, 17, 18, 21, 29-33, 35-50, 53-58, 63, 64, 66, 69, 73, 84, 85, 87, 98, 101, 106-108, 113, 133, 136, 137, 144-154, 157-161, 163-165
Azerbaijani Popular Front (APF), 44, 57, 58

B

bin Laden, Osama, 17
birth rate, 87
Black Sea Economic Cooperation Organization, 7, 138
Black Sea, 7, 8, 17, 21, 22, 23, 29, 32, 33, 41, 48, 49, 52, 61, 64, 69, 84-86, 94, 101, 113, 114, 137, 138
Bolshevik(s), 72, 73, 92, 148, 153
border security, 13
Bulgaria, 63, 118, 119, 156
Bush, President George W., 12

C

Cabinet of Ministers, 58, 65, 86, 123, 127, 130
Canada, 7, 48
Caspian oil deal, 57
Caspian Sea, 33, 43, 46, 48, 56, 61, 69, 85, 113
Caucasus region, 17, 33, 64, 68, 72, 113
censorship, 47, 132
Central Asia, 33, 98, 157
Chechen rebel(s), 17, 20
Chechnya, 11, 17, 84, 123
China, 115, 119, 133, 135
Christianity, 32, 68, 91, 144
civil disorder, 8
civil unrest, 142
civil war, vii, 1, 4, 11, 62, 64-66, 99, 115
Clinton presidency, 12
Clinton, President, 7, 8, 41, 134, 135
coal deposits, 22
commercial relations, 44, 73
Commonwealth of Independent States (CIS), 4, 6, 7, 16, 18, 19, 38, 40, 46, 48, 51, 56, 63, 65, 66, 79, 97, 115, 133, 138, 141, 152, 153, 159
Communist Party of the Soviet Union (CPSU), 96, 153, 155
Communist Youth League, 74
Conference on Security and Cooperation in Europe (CSCE), 7, 36, 40, 46, 55, 66, 83, 153, 159
conscription, 141
Conventional Forces in Europe Treaty (CFE Treaty), 153
coupon (Georgian currency), 8, 52, 106
criminal justice system, 15
Cyrillic, 33
Czechoslovakia, 63, 118, 119, 156

D

decentralization, 74
defense budget, 16
Defense Department (DOD), 12

democracy, 38, 53
democratization, 2, 12, 130
dictatorship, 22, 47, 129
diplomatic isolation, 133
dissident activities, 3
dram, 37, 42, 153

E

economic collapse, vii, 1
economic cooperation, 47, 52, 133
economic corruption, 101
economic growth, 35, 156
economic policy, 118, 129
economic problems, 11
economic reform, 42, 59, 118, 127
education system, 96
election law, 121-124
electoral system, 121, 123, 124
energy crisis, 38, 40, 53, 107
energy needs, 17, 21, 107, 109
environmental problems, 86
ethnic conflict, 6, 16, 35, 36
ethnic crises, 82
ethnic minorities, 3, 6, 32, 49, 74, 76, 123, 136
Ethnic Minorities, 87, 101, 132
European Bank for Reconstruction and Development, 7, 39, 66
European Union (EU), 39, 53
Europe-Asia oil transport corridor, 23
exploratory drilling, 25

F

financial chaos, 33
folk dance, 94
folk music, 94
foreign investment, 12, 35, 50, 53, 56, 99, 100, 119
foreign policy, 5, 12, 15, 38, 45, 59, 134, 136
foreign relations, 65, 84, 133

T

U